VOL 7

근린 생활 시설

NEIGHBORHOOD FACILITY

EDITORIAL NOTE

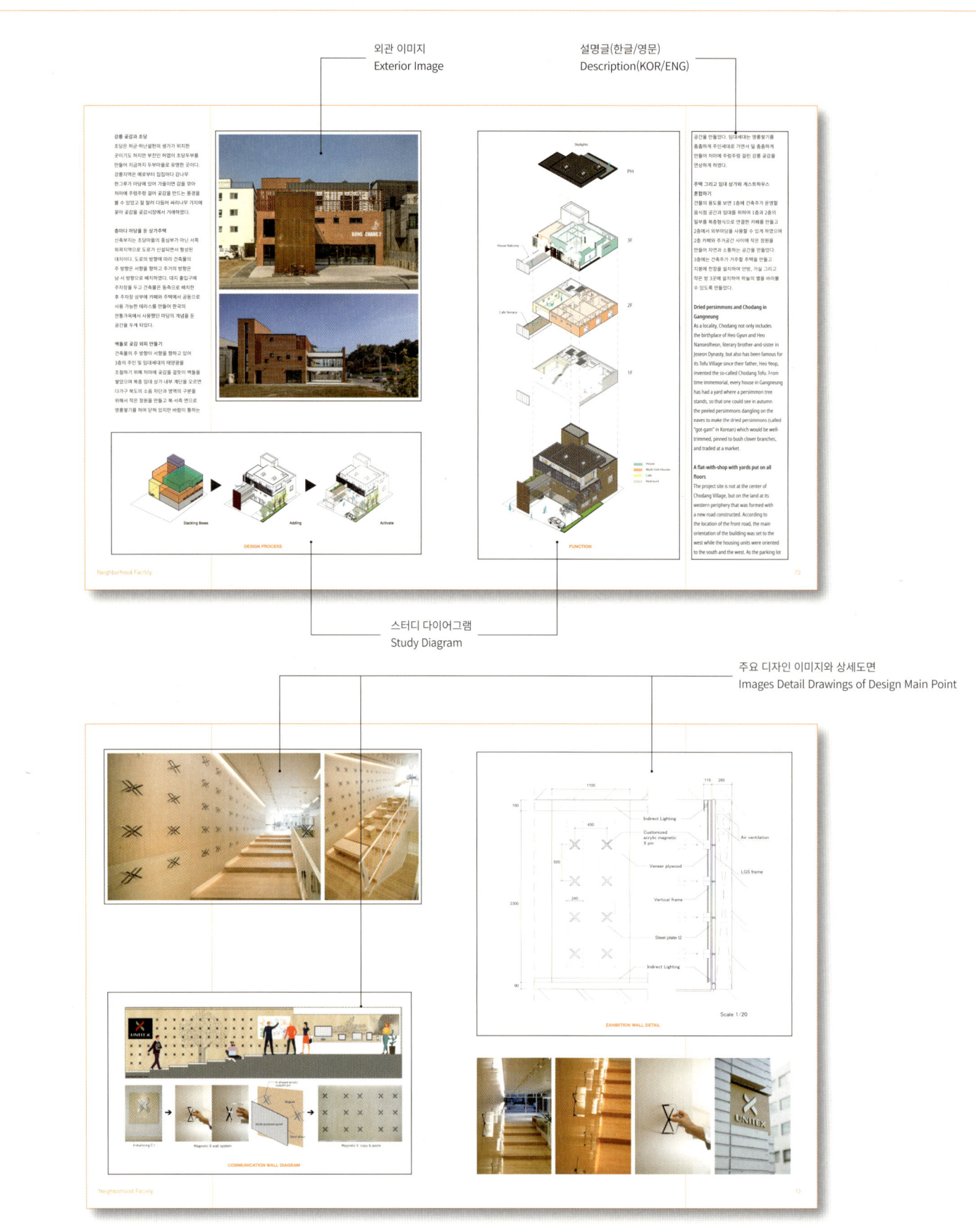

Neighborhood Facility

Neighborhood Facility 7

외관 상세이미지와 관련 상세도면
Facade Detail Image and Related Detail Drawings

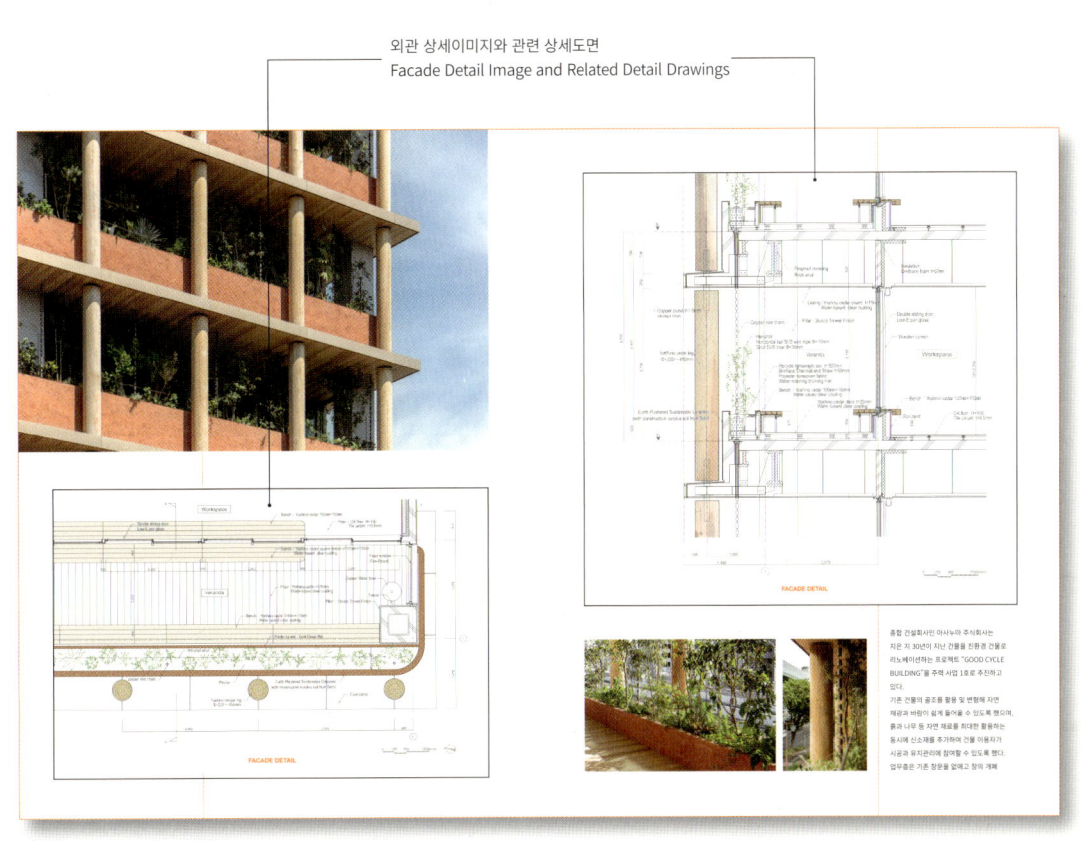

내외부 연관 이미지
Exterior & Interior Images

건축가 소개
Introduction of firm

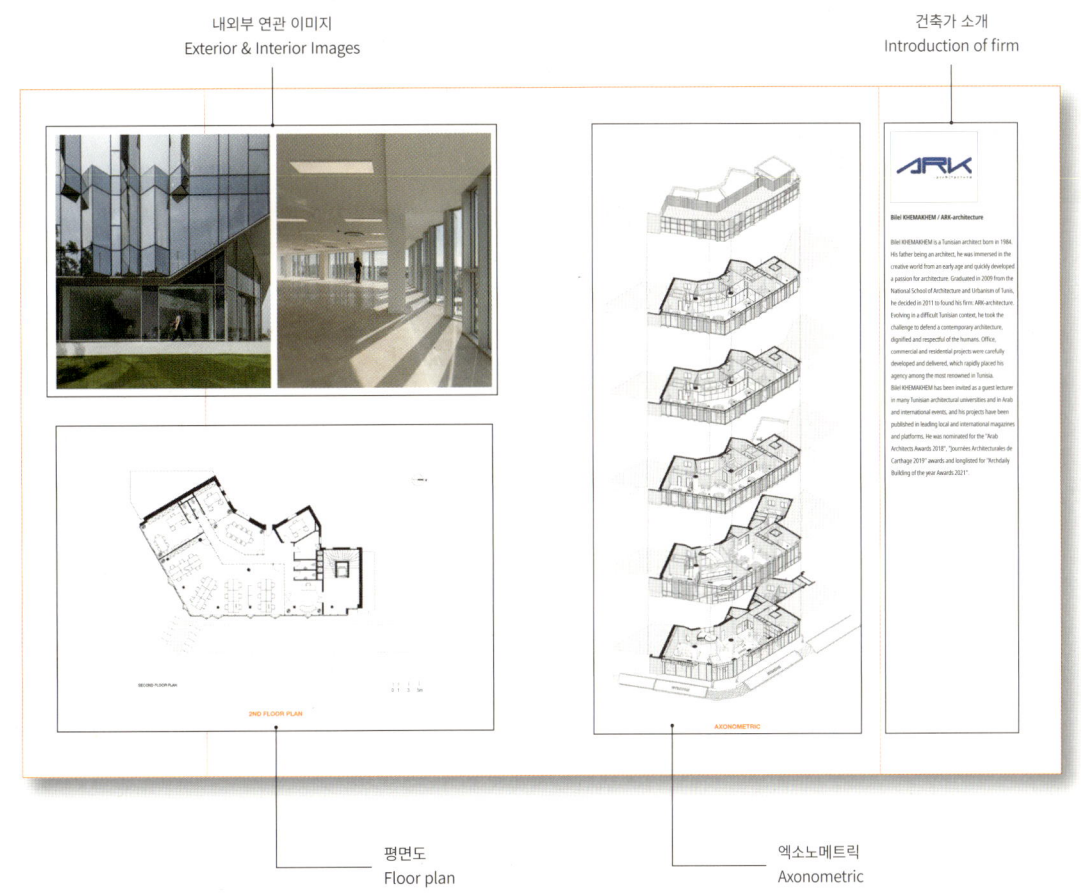

평면도
Floor plan

엑소노메트릭
Axonometric

근린생활시설 7

Contents

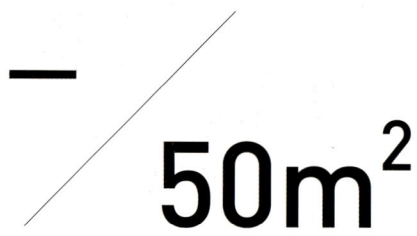
— / 50m²

CAFFÈ DEL POPOLO / 카페 델 포폴로 ESTUDIO RARE	014

TINY FOREST / 작은 숲 정영한 아키텍츠 YOUNGHAN CHUNG ARCHITECTS	022	**RESTAURANT IN OGAWA / 오가와 레스토랑** UENOA ARCHITECTS	056
HANAZONO RESIDENCE "TIME / LIGHT" 하나조노 레지던스 "시간과 빛" JUN MURATA / JAM	034	**JUNG BLDG. / 정빌딩** 엘케이에스에이 건축사사무소 LKSA	066
PI π 3.14 / 옥인동 79-2 근린생활시설 파이3.14 어반아크 URBAN ARK	046	**NIKI BAKERY / 니키 베이커리** KAMAKURA STUDIO	072

Neighborhood Facility

Neighborhood Facility 7

51m² / 100m²

TETUSIN DESIGN RE-USE OFFICE /
재구성을 통해 설계한 새로운 사무실 — 082
YHA ARCHITECTS

MIIKE / 미케 — 098
ihrmk

ACB LIVING / ACB 리빙 — 110
LIGHTHOUSE

MH OFFICE / MH 오피스 — 120
GENDAI SEKKEI CO.,LTD.

ARCHITECTURE AND INTERIOR DESIGN
STUDIO OFFICE / 건축&인테리어 스튜디오 사무실 — 134
Manoj Patel Design Studio

NINE CUBE / 나인큐브 — 146
(주)서원건축사사무소 / SEO WON Architects Cooperation

JAEDONG PROJECT / HALF & HALF BUILDING
재동 프로젝트 / 반반건축 — 156
바운더리스 아키텍츠 / BOUNDARIES ARCHITECTS

근린생활시설 7

101m² / 120m²

AZARA 891 / 아자라 891 — 172
CENTRO CERO

KGA DESIGN STUDIO / KGA 디자인 스튜디오 — 182
Kreis Grennan Architecture

GOMBI / 곰비빌딩 — 214
㈜오엠건축사사무소 omm architects

BIG STAIRS / 빅 스테어스 — 192
오후건축사사무소 + (주)유타건축사사무소
OHOO ARCHITECTURE + UTAA COMPANY

SLEEPING LAB·ARCH / 슬리핑 랩-아치 — 222
Atelier d'More

CUP OF TEA ENSEMBLE / 차 한 잔의 앙상블 — 202
Kraft Architects

OUTDOOR OFFICE / 아웃도어 오피스 — 238
ANDERS BERENSSON ARCHITECTS

Neighborhood Facility

Neighborhood Facility 7

$121m^2$ / $190m^2$

THE FLYING WALLS HOSTEL / 더 플라잉 월 호스텔 — 250
KOMAL DHULIA / DHULIA ARCHITECTURE DESIGN

NGÓI SPACE / 타일 공간 — 258
타일 공간

KAMA-ASA SHOP / 가마아사(釜浅) 상점 — 274
KAMITOPEN CO., LTD.

SUNU BLDG. / 선우빌딩 — 282
엘케이에스에이 건축사사무소 LKSA

W 134 (INOSYS HEADQUARTERS OFFICE) 더블유 일삼사(이노시스 사옥) — 288
(주)오엠엠건축사사무소 omm architects

VIRA II / 비라 II — 296
ALIDOOST AND PARTNERS

PENNANT THONGLOR / 레스토랑 페넌트 통로 — 310
PAD SPACE ARTISAN

근린생활시설 8

191m² / 250m²

SCENERY ALLEY / 시너리 엘리 TEAM_BLDG	014	**X OFFICE_UNITEX HQ** / X 오피스-유니텍스 본사 T2P ARCHITECTS OFFICE & P.O.T LAB	054
HARMAY FANG / 하메이 팡 AIM ARCHITECTURE	030	**DRIED PERSIMMON HOUSE** / 초당동 곶감집 건축사사무소 예인 ARCHITECTURE STUDIO YEIN	070
VERTICAL URBAN / 버티컬 어반 건축사사무소 H2L ARCHITECTS H2L	044	**THE STRANGERS** / 더 스트레인저스 WUTOPIA LAB	082

Neighborhood Facility

Neighborhood Facility 8

251m² / 300m²

ÖTZTAL TOURISMUS / 외츠탈 관광안내소 OBERMOSER + PARTNER ARCHITEKTEN ZT GMBH — 096	**PALETTE / 팔레트** KOSUKE BANDO ARCHITECTS + NANOMETER ARCHITECTURE — 146
EST62 / EST62 SILP ARCHITECTS — 116	**URBAN SQUARE / 어반스퀘어** (주)건축사사무소 플랜 PLAN ARCHITECTS OFFICE — 166
DENTAL PRACTICE IN KAMPEN / **캄펜의 치과진료소** BURO/S ARCHITECTS — 134	**FRESH CORPORATION OFFICE / 후레쉬상사 사옥** 무아키텍츠 건축사사무소 MOO ARCHITECTS — 178

근린생활시설 8

301m² / 450m²

GOOD CYCLE BUILDING 001 / 굿 사이클 빌딩 NORI ARCHITECTS + ASANUMA CORPORATION	192
ANH COFFEE ROASTERY / 안 커피 로스터리 RED5	206
ETRATON / 에트라톤 리모델링 프로젝트 ARK-architecture	220
RESTAURANT OF METASEQUOIA GROVE / **메타세쿼이아 숲 레스토랑** GOA(GROUP OF ARCHITECTS)	228
THE BOILER ROOM / 보일러룸 ALTA ARCHITECTES URBANISTES	238

Neighborhood Facility

Neighborhood Facility 8

451m² / —

REMED HEADQUARTERS / 위례 리메드 사옥 '산해' 256 투닷건축사사무소 주식회사 TODOT ARCHITECTS AND PARTNERS	**JIANLI ART CENTER / 지안리 아트센터** 298 GOA (GROUP OF ARCHITECTS)
TURNAROUND OF AISAIKAN / 268 **턴어라운드 오브 아이사이칸** B2AARCHITECTS	**THE EAVES / 디 이브스** 312 BUREAU^PROBERTS
HHT COFFEE / HHT 커피 286 P.I ARCHITECTS	**COR MEUM / 꼬르메움** 326 (주)건축사사무소가로 KARO ARCHITECTS

근린생활시설 7

— 50m²

NEIGHBORHOOD FACILITY 7

CAFFÈ DEL POPOLO / 카페 델 포폴로
ESTUDIO RARE

TINY FOREST / 작은 숲
정영한 아키텍츠
YOUNGHAN CHUNG ARCHITECTS

HANAZONO RESIDENCE "TIME / LIGHT"
하나조나 레지던스 "시간과 빛"
JUN MURATA / JAM

PI π 3.14 / 옥인동 79-2 근린생활시설 파이3.14
어반아크 URBAN ARK

RESTAURANT IN OGAWA / 오가와 레스토랑
UENOA ARCHITECTS

JUNG BLDG. / 정빌딩
엘케이에스에이 건축사사무소 LKSA

NIKI BAKERY / 니키 베이커리
KAMAKURA STUDIO

Location
Córdoba, Argentina
Use
Café
Site area
4.03m^2
Built area
4.03m^2
Total floor area
11.45m^2
Floors
3F

Exterior finish
Fiberglass, Metal
Interior finish
CNC Carved Plywood
Project architect
Agustín Willnecker, Mateo Unamuno, Ivan Ferrero, María Belén Marinelli
Photographer
Estudio RARE

Caffè del Popolo

카페 델 포폴로

ESTUDIO RARE

A PROSTHESIS.

Cafe del Popolo Caffè del Popolo is one of those experiments that arise in those spaces that are often understood as unusable or obsolete in the city. Work was carried out in a triangular void between two blocks of flats in the Nueva Córdoba neighborhood. The perception of compression suffered by its context triggered the initial exploration for the design of this enclosure, which from the beginning was conceived as a device capable of fitting into that place, such as an "urban prosthesis". It proposes a way to enter a dense urban environment and act in it and from it towards the city.

THE FLUIDITY.

The insertion in such a small space and surrounded by large buildings, printed the idea of compression in our material models and in the different explorations carried out. Perhaps it is also the liquid, dynamic and fluid perception of that corner and of the current times that ordered the formal action resulting from the proposal. The fluidity in terms of use also determined design actions, and how to generate a space as open as possible to the street in its zero plane.

MATERIAL EXPLORATION.

A metallic base was created, with guillotine openings, capable of achieving a maximum opening towards the street without neglecting the functional requirements. In the interior space, through digital fabrication, a routed and later lacquered wood cladding was achieved, inspired both in its color and in its formal composition, in the coffee plant. For the upper levels, an existing metal structure was used, from which five panels made of fiberglass were taken. These panels are the ones that protect two levels towards the inside: One with a bathroom and the other with a deposit.

Neighborhood Facility

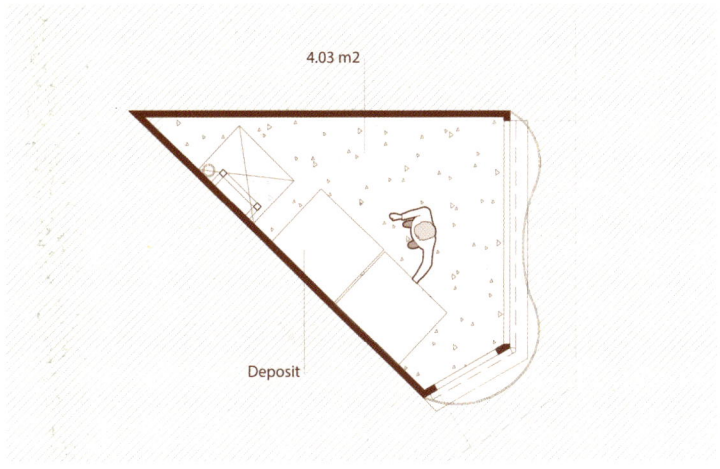

LEVEL 2

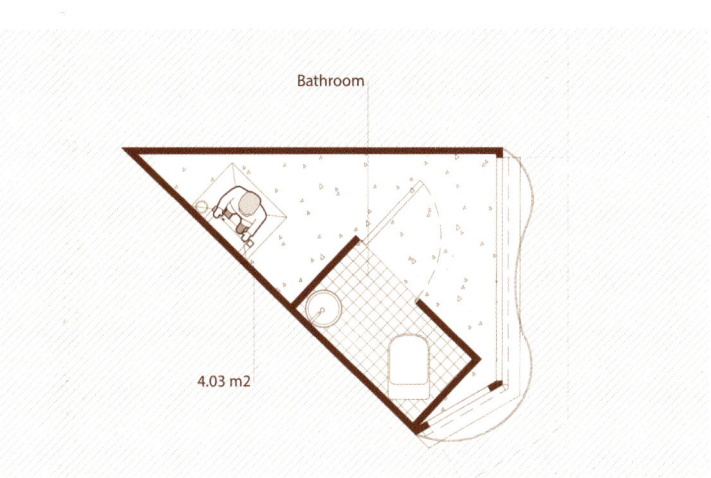

LEVEL 1

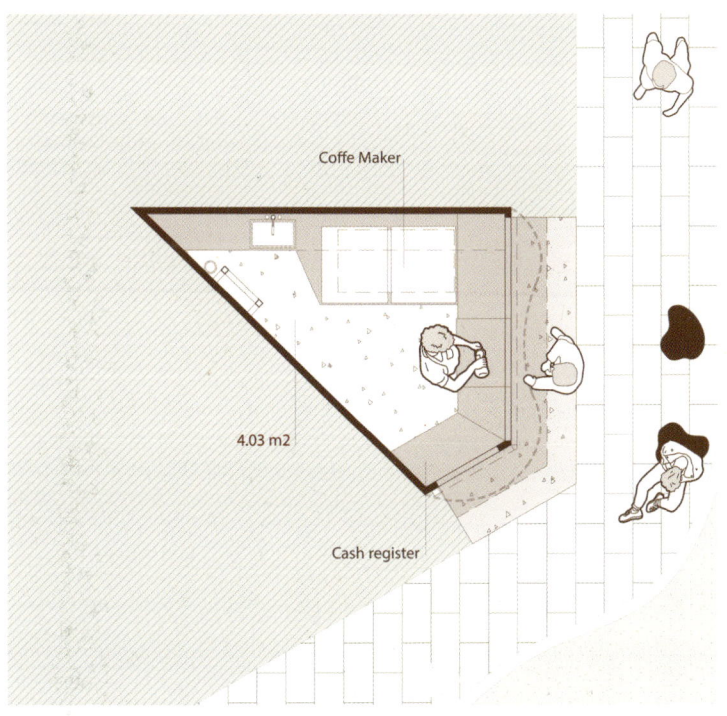

FLOOR PLAN

STREET.

Using the street is a fundamental premise, and not only because there is a reduced interior space, but also because there was an intention to provide that transit space with a pause, a space for contemplation, a possible use, taking advantage of what the urban landscape Cordobés offers in that area, such as the residential building of the architect Ignacio Togo Diaz, the Cultural Walk of "El Buen Pastor", the Church of the "Capuchinos and the Building of the architect Osvaldo Pons. An interior/exterior support surface was created on the cafe's façade, a platform where two mobile metal stools are located, and a fixed urban furniture (Rarx Bench) made of concrete and metal was developed, located on the curb of the sidewalk. The Rarx benches finish symbolically synthesizing the generating idea of the project, since they also contemplate material, formal and use reflections in relation to public space.

삽입 건물

카페 델 포폴로(Caffè del Popolo)는 도시에서 자주 활용이 불가하거나 쓸모 없는 것으로 여겨지는 공간에서 출발한 실험 중 하나이다.

작업은 누에바 코르도바(Nueva Córdoba) 인근 아파트 두 동 사이 삼각형 모양의 빈 공간에서 진행했다. 이런 주변 환경에서 오는 압박을 인지하면서부터 아파트로 둘러 싸인 공간에 대한 디자인 연구를 시작했으며, 이는 "도시의 삽입 건물"처럼 처음부터 아파트 사이의 빈공간에 어울릴 수 있는 하나의 방법으로 보았다.

이런 디자인은 밀집된 도시 환경으로 들어가 그 공간 안에서 움직이고 동시에 도시를 향하는 방법을 제안한다.

유연성

이렇게 작은 공간 안에 큰 건물로 둘러싸인 삽입된 건물은 자재 모델과 다양한 탐색을 통해 압축에 대한 아이디어를 드러낸다. 해당 제안에서 시작된 공식적 조치를 가능하게 한 건 아마도 아파트 사이의 코너와 현 시대에 대한 유동적이고 역동적이며 유연한 인식일 것이다. 사용 측면에서의 유연성은 설계 조치와 더불어 가급적 거리로 이어지는 개방된 평면적 공간을 결정한다.

재료 탐색

기능적 요구사항을 존중하여 최대한 거리를 향해 개방된 길로틴 출입구가 있는 금속 베이스를 만들었다. 내부 공간의 색상과 전반적인 구성은 커피 나무에서 영감을 얻은 목재 클래딩을 디지털 패브리케이션 방식으로 레이아웃을 정하고 광택제를 발랐다. 상부 층은 기존의 메탈 구조(다섯 개 유리 섬유 패널 구조)를 적용했다. 이들 패널의 용도는 내부를 향한 두 개 층, 즉 화장실층과 저장고층을 보호하는 것이다.

거리

기본 전제는 거리의 활용이다. 그 이유는 내부 공간이 좁을 뿐 아니라 코르도베스(Cordobés) 도시 경관이 주는 장점을 살려 잠시 멈출 수 있는

BEFORE

환승의 공간, 사색을 위한 공간, 실용적인 공간을 제공하려는 의도 때문이다. 이러한 건축물로는 건축가 이그나시오 토고 디아즈(Ignacio Togo Diaz)의 주거용 빌딩, "엘 부엔 목사(El Buen Pastor)"의 문화 산책, "카푸치노"의 교회, 건축가 오스발도 폰스(Osvaldo Pons)의 빌딩 등이 있다.

카페 정면에는 이동식 메탈 스툴 두 개가 놓인 플랫폼으로 건물 내외부 지지면을 조성하고, 보도 연석에는 콘크리트와 메탈 소재의 고정형 어반 스타일 가구(랙스 벤치: Rarx Bench)를 설치했다. 랙스 벤치는 재료, 형식, 사용에 공용 공간과의 연결을 고려해야 하므로 본 프로젝트 아이디어 개발에 대한 상징적 통합을 완성한다.

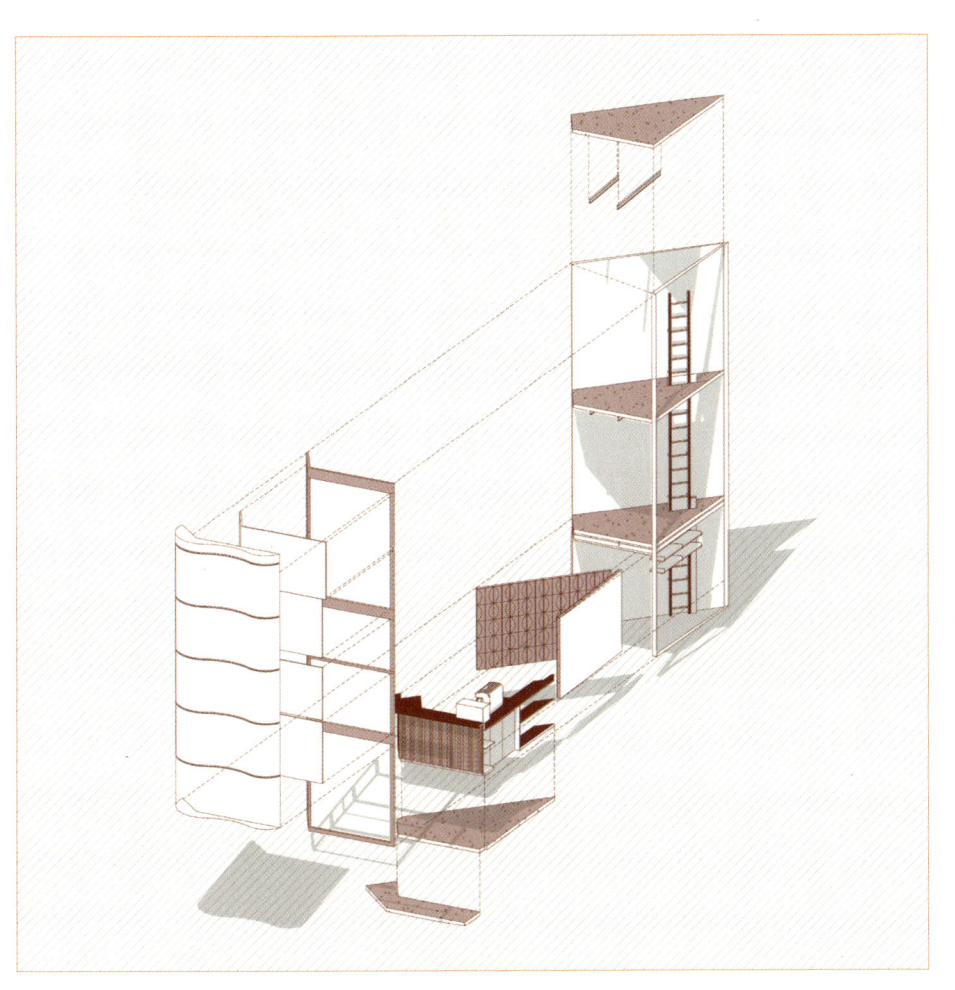

AXONOMETRIC

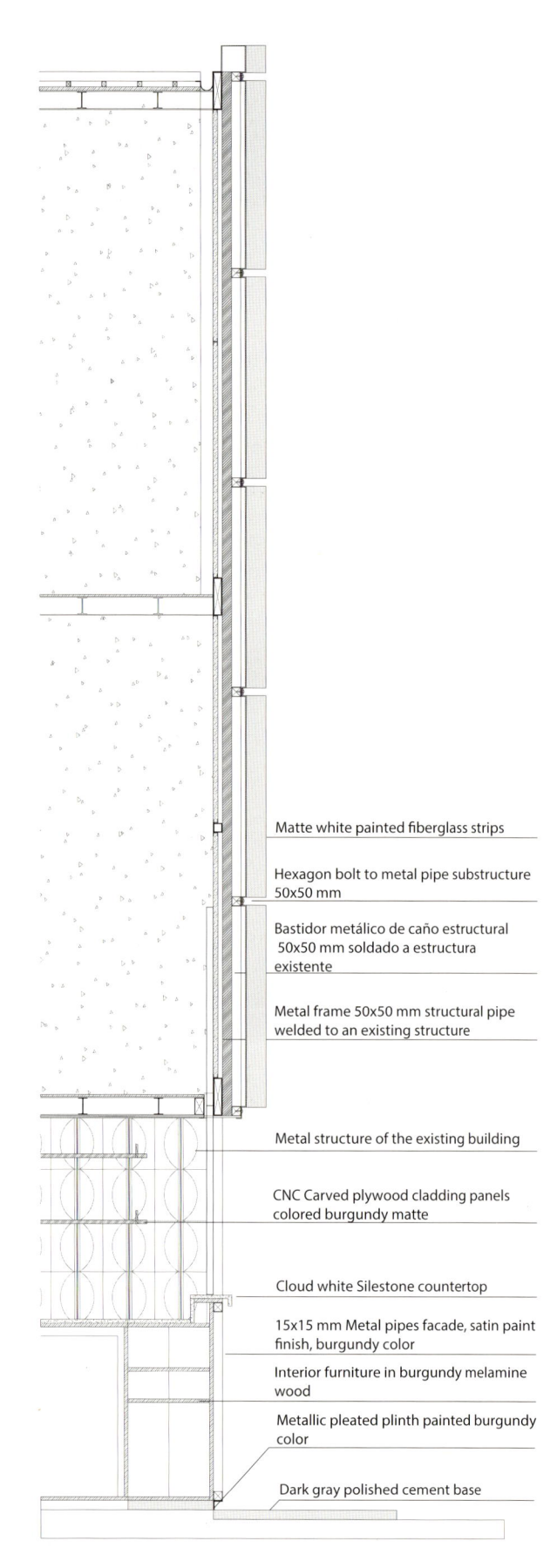
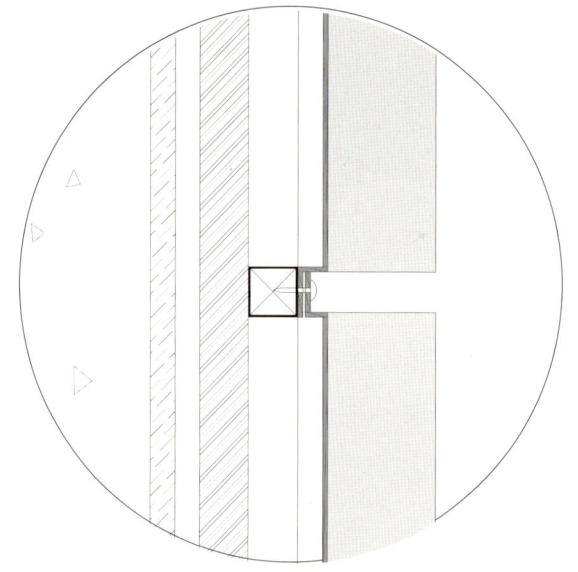

- Matte white painted fiberglass strips
- Hexagon bolt to metal pipe substructure 50x50 mm
- Bastidor metálico de caño estructural 50x50 mm soldado a estructura existente
- Metal frame 50x50 mm structural pipe welded to an existing structure
- Metal structure of the existing building
- CNC Carved plywood cladding panels colored burgundy matte
- Cloud white Silestone countertop
- 15x15 mm Metal pipes facade, satin paint finish, burgundy color
- Interior furniture in burgundy melamine wood
- Metallic pleated plinth painted burgundy color
- Dark gray polished cement base

DETAIL

Neighborhood Facility

RARE

Estudio RARE

It is a collective formed by architects, designers and artists founded by Mateo Unamuno, Ivan Ferrero and Agustín Willnecker, based in Córdoba, Argentina. We consider as a contemporary condition to think architecture through art and art through architecture. We work on a design where both are understood as a whole, we approach our design and creative processes from experimentation, a tool that allows us to "discover" without wanting to "invent" and, in turn, generate unexpected results.

All our works have an explanation because we understand the conceptual search as an essential element to give meaning to creation.

We think of design as a situated and circumstantial fact, its closest context is essential to belong to a culture, history and society in particular, but its confrontation with the world, also gives it an irradiation of the contemporary that reflects the spirit of the time.

We work in an interdisciplinary and multi-scalar way, finding in design the channel to express ourselves and transform reality. We consider teamwork as a potential and having fun on a daily basis, a necessity.

\>> estudiorare.com

Neighborhood Facility

Location
17 Bukchon-ro 1-gil, Jongno-gu, Seoul, Republic of Korea
Use
Neighborhood living facilities
Site area
58.83m²
Built area
31.87m²
Total floor area
71.37m²
Floors
2F

Exterior finish
Aluminum design panel, Ash thermowood.
Interior finish
Birch plywood on transparent varnish fin.
Project architect
Chung Younghan
Construction
TCM Global
Photographer
Yoon Joonhwan

Tiny Forest

작은 숲

정영한 아키텍츠
YOUNGHAN CHUNG ARCHITECTS

작은 건축

반듯한 택지개발로 정형화된 필지와 달리 한옥들이 즐비했던 과거의 시간을 고스란히 담고 있는 도심 속 소규모 필지는 우리에겐 그리 낯설지 않다. 태생적으로 작은 필지에서 작은 건축은 당연한 결과일 텐데 일본의 협소 건축과 우리의 작은 건축은 무엇이 다를까? 아마도 불필요한 공간들을 최대한 제거하고 작은 공간에 맞는 근본적인 거주방식이 오랜 삶 속에 문화로 자리 잡은 점이다. '일어서면 반조. 누우면 다다미 한조' 라는 그들의 속담에서 보듯 다다미 4조 반의 공간성을 통해 그 작음에 대한 이질감이 우리에 비해 덜할 수밖에 없지 않을까? 그와 달리 우리에겐 경제적 이유가 협소 건축을 하는 데 있어 첫 번째로 작동한다. 지가상승 탓에 소규모의 필지를 찾고 건축가들은 그 규모에 맞는 건축 유형을 탐색하고 사용자들 역시 스몰 라이프에 적응하고 있다. 그러나 공간을 구성하는 실을 유지하기 위해 그 크기만을 줄이는 것엔 반드시 한계가 따른다. 작은 건축은 작아야 할 필연성이 구조와 공간에 모두 반영되어야 한다.

주택 안에 있는 취미 공간이나 서재와 같은 공간들은 주거라는 틀 안에서 점차 느슨해져 본연의 기능의 힘을 잃어가고 오히려 주택에서 벗어나 자신을 위한 소우주와 같은 공간을 경험해 보고자 하는 욕망은 현대에 사는 우리 모두에게 절실하다. 본 건물은 주거 안에서 부수적 기능을 가진 서재나 손님을 맞이하는 기능을 외부로 분리하여 독립된 도시 속 작은 사랑방을 만들고자 했다. 오랜 시간 대학에서 학생들을 가르치시고 은퇴하신 노년의 건축주는 공통관심사를 가진 지인들과 가끔 와인과 함께 읽은 책들에 관해 자유로운 대화를 하거나 인왕산을 바라보며 자신을 고요히 마주하며 힐링할 수 있을 다양한 잠재적 가능성의 작은 건축을 기대하였다. 늘 나의 건축에서 공간의 완결은 물리적 상태를 만들고 빈집을 떠나는 건축가의 몫이 아닌 사용자에 의해 완결된다고 생각해왔다. 규모의 문제가 아닌 그 공간의 기능과 쓰임이 큰 규모의 건축 이상으로 사용자에 의해 다양하게 번역될

SITE PLAN

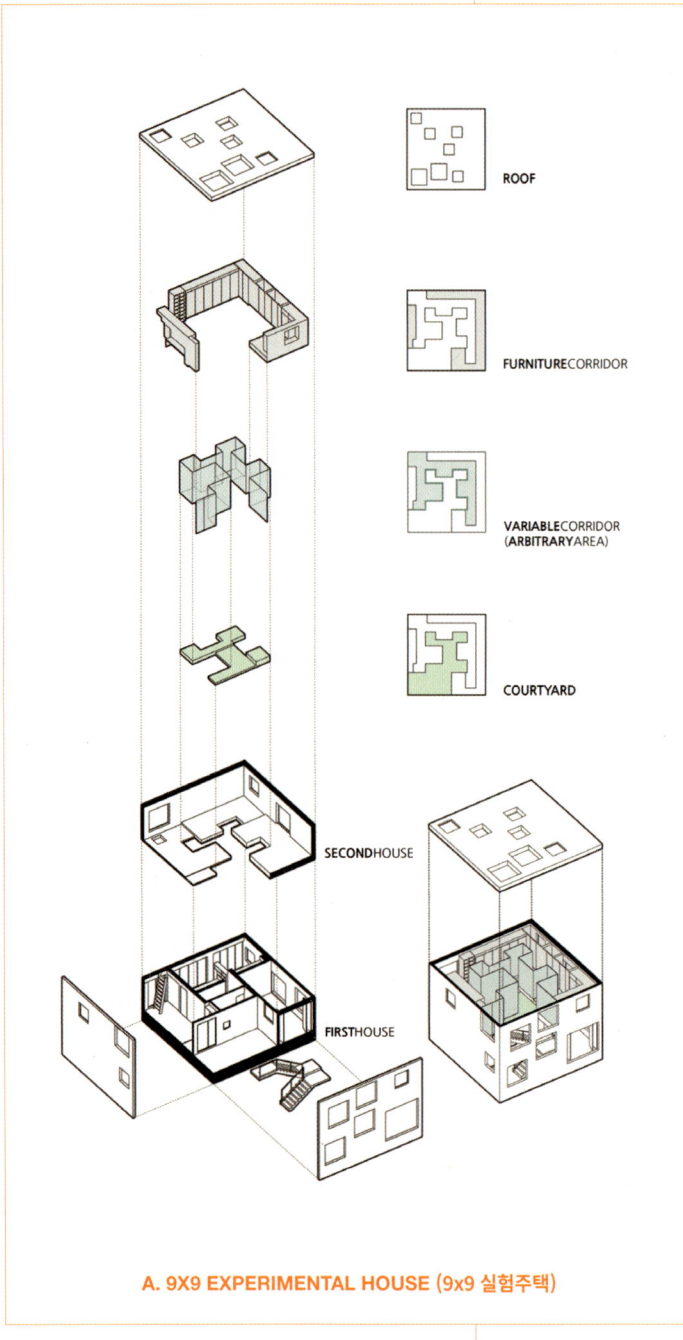

A. 9X9 EXPERIMENTAL HOUSE (9x9 실험주택)

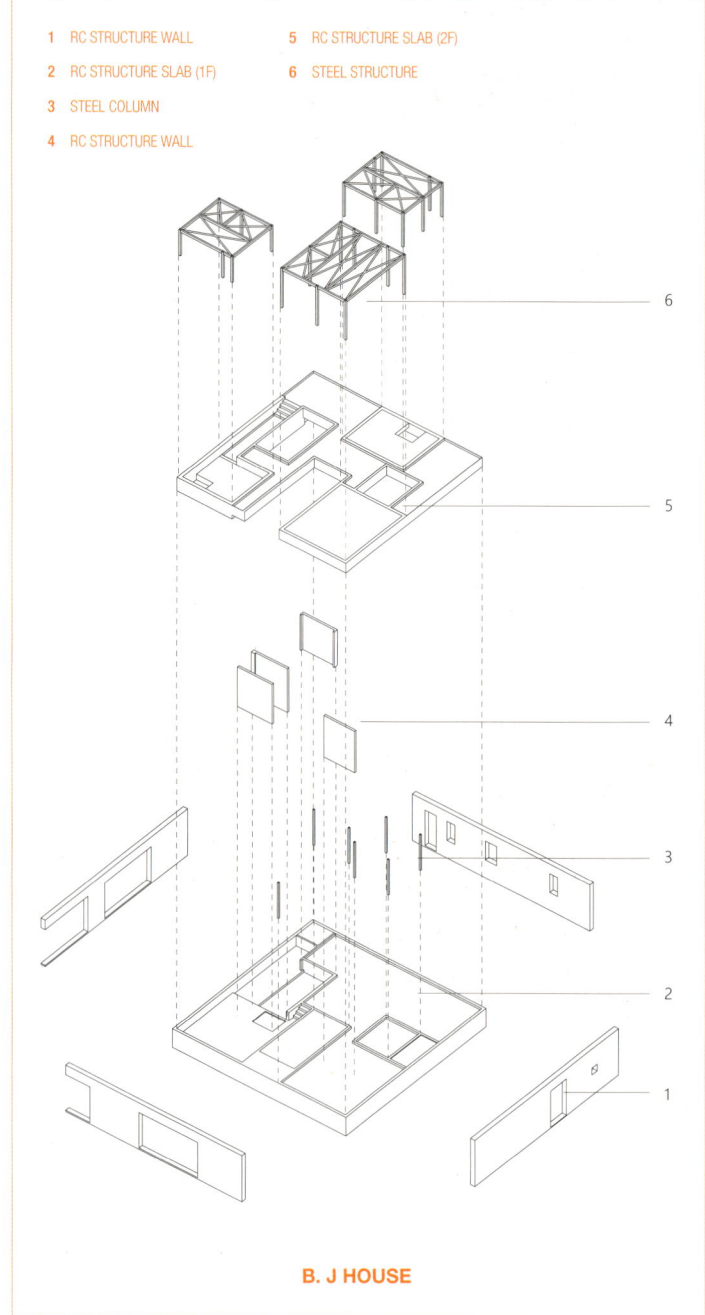

B. J HOUSE

가능성을 품는다면 시간의 변화를 거스르지 않고 더 단단히 견디어 낼 수 있는 것이 작은 건축이라 생각한다.

작은 구조

공간을 위한 구조, 구조에 의한 공간을 스스로 경계하면서 양자가 공존하며 동시성을 획득할 수 있도록 초기 기획 단계부터 설정한다. 자칫 관습적인 구조 방식으로 인해 다양한 삶의 방식이 투영된 사용자의 안무(按舞, choreography)를 엄격히 규정된 틀 안으로 가둘 수 있는 위험이 늘 도사리기 때문이다.

내부의 내력벽이나 주요 기둥을 제거하고 다공(POROUS)으로 구성된 4면의 외벽만이 구조벽으로 완성된 작은 주택 <9X9실험주택>이나 3개의 중정 외곽의 최소 스틸 기둥(100*100)으로 오픈 플랜을 만들었던 주택 <J HOUSE> 그리고 150*150의 부재를 주 기둥으로 과감히 처리하며 부유하는 슬라브를 만들었던 주택 <정은설>을 통해 구조가 공간을 규정하지 않는 것에 관심을 두고 있다. 본 프로젝트 역시 작은 건축에 걸맞은 구조가 요구되었다. 미세한 각도로 틀어지거나 뻗어 나간 매스들을 보면 마치 컨테이너와 같은 레디메이드(Ready-made) 형식으로 보이지만 실제로는 세장한 비율의 철골 부재와 틀어진 두 개의 1,2층 매스를 위한 전이보(RANSFER BEAM)구간을 설정한 철골조이다. 골목길로 향해 약 3m 정도 뻗어나간 작은 매스는 지름 89.1의 CFT(Cement filled tube) 기둥을 적용하여 최대한 부유한 구조로 만들고자 했다.

Neighborhood Facility

1 STEEL COLUMN
2 ALUMINUM PANEL
3 STEEL BRACE
4 STEEL BEAM
5 SLAB
6 ALUMINUM LOUVER
7 SKYLIGHT

1 ALUMINUM DESIGN PANEL
2 CFT(Ø 89.1) & STEEL BRACE
3 STEEL COLUMN & STEEL BRACE
4 STEEL BEAM
5 ALUMINUM SHEET
6 TRANSFER BEAM
7 FLAT DECK SLAB
8 SQUARE PIPE
9 EXPANDED METAL FENCE
   (SCHEDULED CONSTRUCTION)

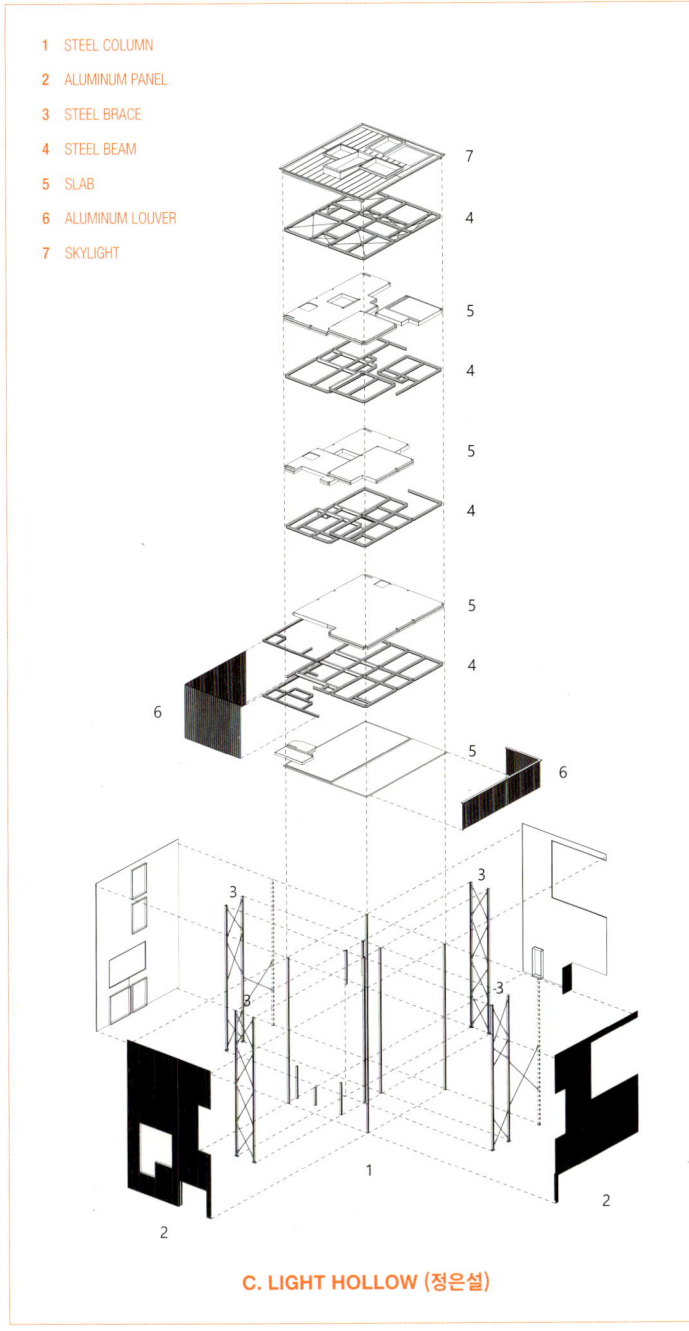

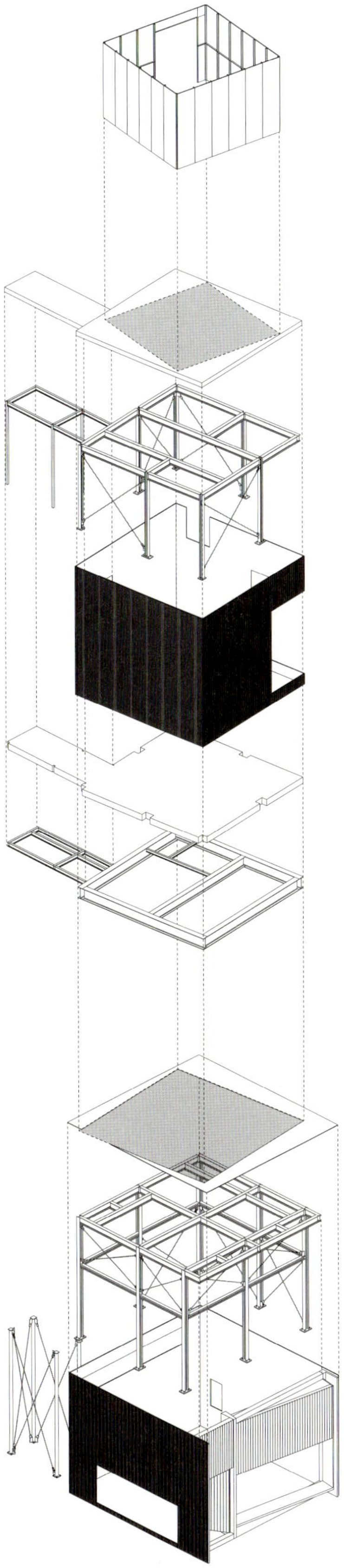

C. LIGHT HOLLOW (정은설)

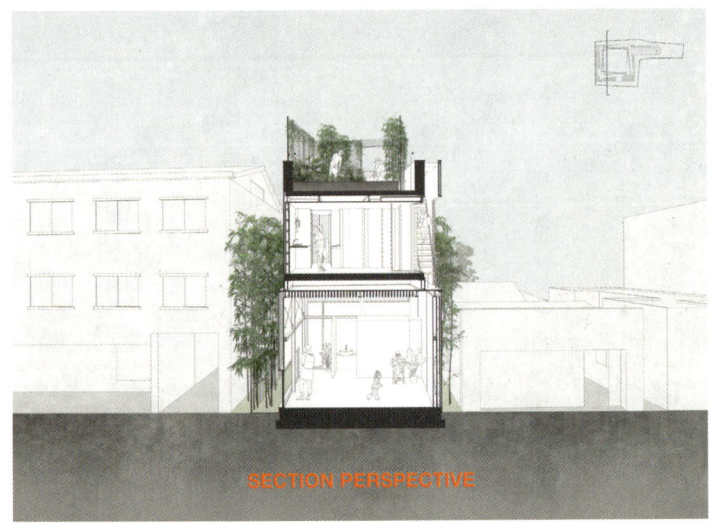

SECTION PERSPECTIVE

D. TINY FOREST (작은 숲)

Small Architecture

In contrast to uniform lots in square housing site development, small lots in the city center that still retain the essence of Hanok, traditional Korean houses, are not uncommon. Small structures on small lots are the obvious outcome. What distinguishes Japan's narrow architecture from our small architecture? Perhaps it is the fact that the fundamental approach of removing as many unnecessary spaces as possible and living in a compact space has become a culture over time. As seen in their adage, "A half piece of tadami for standing, one piece for lying down," it is probably natural that their disfavor of small space is less than ours, considering their spatiality of four and a half tatami mats? On the other hand, economic considerations take precedence in our small architecture. Due to rising land prices, architects are seeking small lots and, consequently, building types suitable for the scale, while users are also adapting to the small lifestyle. However, reducing the size must ultimately reach a limit in order to retain spatial practicality. The inevitable nature of small architecture should be reflected in both structure and space.

Spaces for hobbies and study in a house gradually shrink within the confines of housing and eventually cease to serve their original purpose. Instead, people in modern times have a greater urge to leave the house and explore a space separated from the home. I intended to create an independent small reception in the city by bringing out the ancillary functions of a study or guest welcome from a living space to the outside of the building. The elderly client, who had retired after many years of teaching students at a university, wished to have a small building with lots of possibilities where he could invite friends and discuss common interests about books or to calmly gaze at

Neighborhood Facility

the Inwangsan Mountain over a glass of wine. I have always thought that the users of buildings should complete the spaces rather than the architect, who only produces the physical spaces and leaves the structures empty. I genuinely think that the purpose and use of the space are not a matter of scale, yet small architecture seems more resilient to the ravages of time than large-scale architecture, provided it allows for the user to create their own ambience.

Small Structure
While being careful to avoid the structure being biased toward space and the space being biased toward structure, the initial planning stage is designed to accomplish the coexistence of both. The reason for this is that there is always a risk that the users'

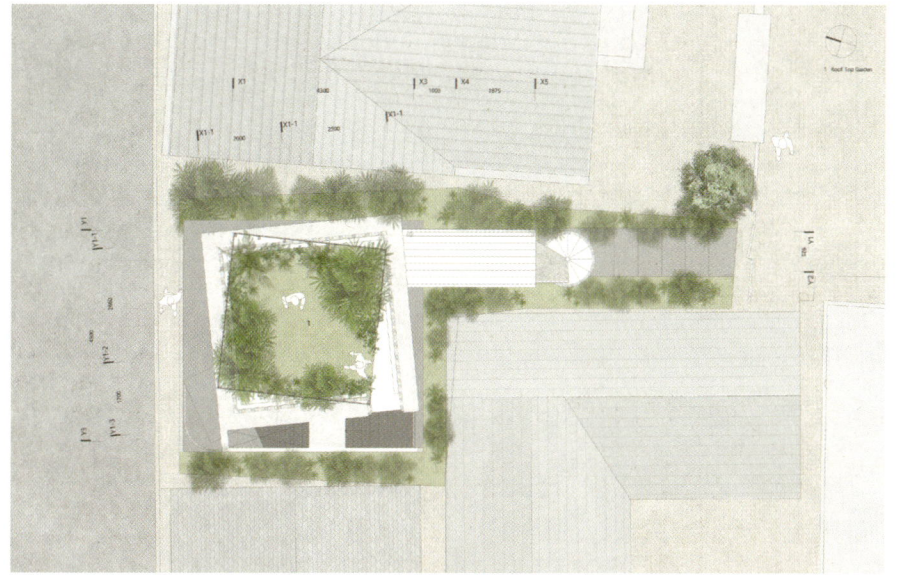

ROOF FLOOR PLAN

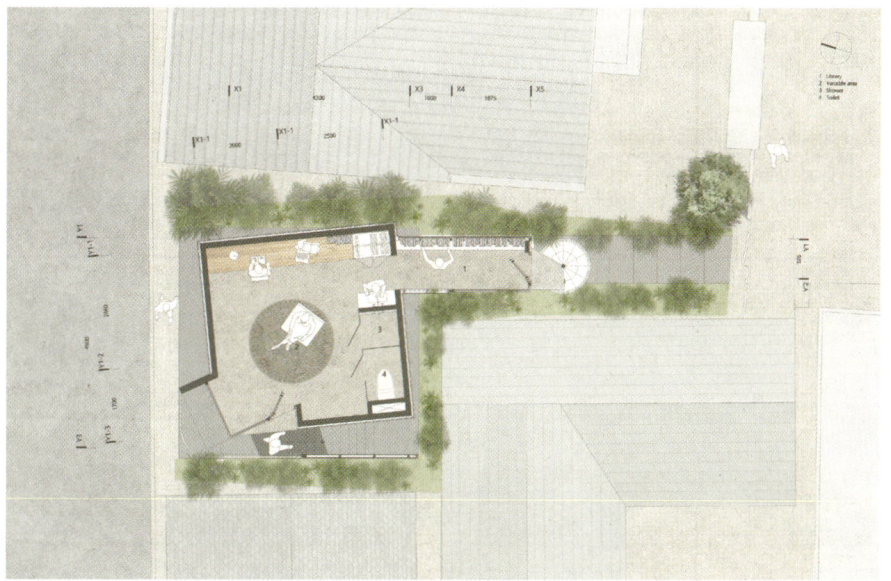

2ND FLOOR PLAN

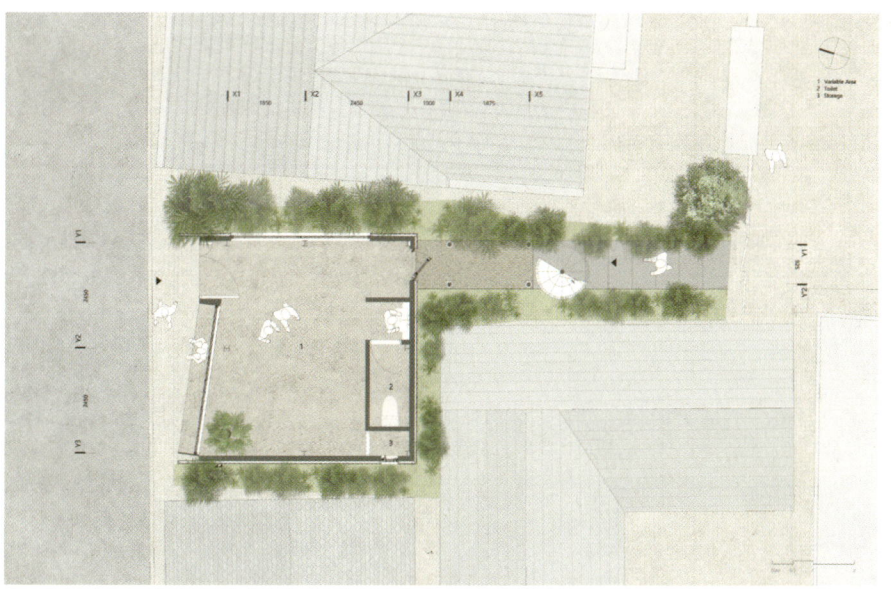

1ST FLOOR PLAN

choreographies, which represent many different lifestyles, can be constrained within the strictly specified framework of the conventional structure method.

The "9x9 Experimental House," a small house with only four porous outer walls after removing the internal walls and main columns; the "J House", a house with an open floor plan made of minimum steel pillars (100 x 100) around three courtyards; and the "Light Hollow", a house that audaciously treats 150 x 150 members as main columns with floating slabs, are all interested in the theme that the structure does not define the space. This project also required a structure appropriate for a small building. Although the masses appear to be ready-made type, like containers, because of the angular extensions and slight twist, but in fact, it is a steel frame with the slender steel members and transfer beam section for the first and second floors' twisted masses. The small mass, which extends roughly 3 meters in the direction of the alleyway, was designed using CFT (cement-filled tube) columns with a diameter of 89.1 to make it as floating as possible.

ELEVATION

Neighborhood Facility

정영한 / 정영한 아키텍츠
Chung Younghan / Younghan Chung Architects

건축가 정영한은 한양대학교 대학원 건축과를 졸업하였다. 2002년 설립된 이후 현재까지 과밀하고 획일화된 도시 풍경 틈에서 새로운 주거 유형을 탐색하는 동시에 다양한 현상과의 관계를 통하여 실험적이고 창의적인 프로세스를 기반으로 한 설계수법을 연구하고 있다. 2013년부터 장기 기획전시인 〈최소의 집〉의 총괄 전시기획을 맡아 진행 중에 있으며 2016년에는 문화체육부 주체의 〈새로운 주거방식의 조각들: 한국현대사회의 도시주거〉전의 초대작가로 선발되어 싱가포르 국립대학에서 전시를 하였다.

대표작으로 〈체화의 풍경〉은 2013년 서울시 건축상을 〈6X6주택, 2014〉으로 김수근 문화재단에서 수여하는 '김수근 프리뷰 어워드'를 수상하였고 〈다섯 그루 나무, 2015〉 한국 건축가 협회상 올해의 베스트 7으로 협회상과 2016년에는 부산다운 건축상을 동시에 수상하였다. 〈물 위의 방, 2018(미국)〉은 THE ARCHITECTURE MASTERPRIZE 주거부문을 수상하고 2020년에는 시카고 아테네움 건축 디자인 박물관(The Chicago Athenaeum)과 유럽 건축예술디자인도시 연구센터가 선정하는 올해의 국제건축상(the International Architecture 2020)에 선정되었다.
>> http://archiholic.com

SECTION PERSPECTIVE

Neighborhood Facility

	Location Osaka, Japan **Use** Accommodation(Guest house) **Site area** 61.94m² **Built area** 40.00m² **Total floor area** 77.00m² **Floors** 2F	**Exterior finish** Acrylic resin painting, ceramic siding **Interior finish** Non-combustible wallpaper **Structure** Timber Frame **Architect** Jun Murata / JAM **Structure Design** Prenda Architect Office **Photographer** JAM

Hanazono Residence "Time / Light"

하나조나 레지던스 "시간과 빛"

JUN MURATA / JAM

This is a guesthouse located in an area with a lot of small houses, close to the center of Osaka. It was required to have a highly adaptable space that is easy to use and comfortable to use, assuming the use of various races and generations visiting here. In addition, the client required that space which has a small court, light and wind, and Zen element which guest can experience Japan.

There was grass on the south side of the site, which was rare in an overcrowded cityscape. This adjacent open space was regarded as a positive surplus space and was a clue in design.
In the gable-shaped volume planned to the limit of the size of the site, a small court of 1.8 meters square was arranged so as to penetrate horizontally toward the open space and vertically toward the sky.
The vertical aluminum screen installed at the boundary between the court and the open space ensures privacy from the surrounding environment and plays an important role in security.

Every rooms and corridors face the court to disperse the light from it. The kitchen, the sanitary and public space were arranged on the 1st floor. There are two guest rooms on both sides of the court on the 2nd floor. The vertical screen composed of aluminum unequal-angle steel of 4 meters length is laid out with the blade side facing inward and the end side facing outward. We designed to open the line of sight from the inside of rooms and the court toward the open space while blocking the line of sight from the outside. By arranging angle steels densely for each, the surrounding environment are cut out in strips, and fragmented. Guests would be able to get a glimpse of the surrounding situation and feel the connection with the environment even

Neighborhood Facility

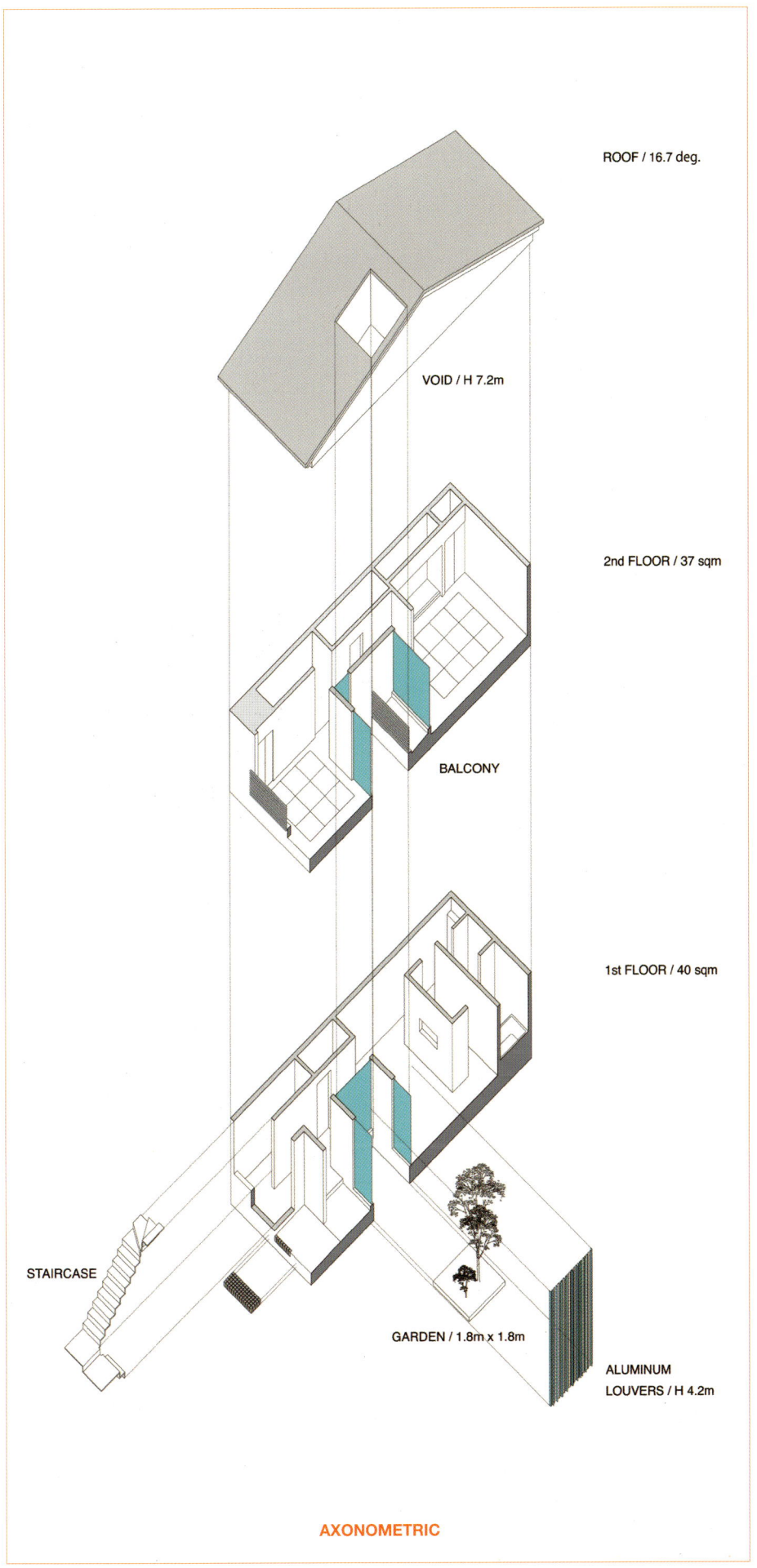

AXONOMETRIC

though it is closed.

Several linear lights transform the density and cast an impressive shadow on the black stones laid in the court. The court extends horizontally to the open space, giving a feeling of depth wider than the actual size. There is no obstruction between the upper part of the aluminum screen and the ceiling beam on the second floor, and the view of the sky spreads horizontally.

Looking directly above the court where the high and low trees are planted, a cut-out sky appears. Guests can enjoy various changes in the sky through the branches and leaves swaying in the wind. The shadows of the trees and screens are projected sharply inside the white interior. The activities of nature are engraved in a modest and neutral space with the passage of time. We aimed to create a space full of light and wind so that guests could remember their stay more clearly.

Neighborhood Facility

이 건축물은 소형 주택이 즐비한 오사카 중심부 근처에 위치한 게스트하우스이다. 클라이언트는 다양한 인종과 세대가 이 곳을 찾는다는 가정하에 게스트가 쉽고 편리하게 이용할 수 있는 매우 유연한 공간이 되길 바랐다. 또한 작은 안뜰에 빛과 바람, 선(Zen, 일본식 불교) 요소를 두어 게스트가 일본을 경험하길 원했다.

게스트하우스 남쪽에는 혼잡한 도시에서는 보기 드문 풀밭이 펼쳐져 있었다. 게스트하우스에 인접한 공개공지는 양(positive)의 잉여 공간으로서 디자인의 실마리가 되었다. 부지 크기의 한계 때문에 계획한 박공지붕 모양의 공간 안에 있는 작은 1.8m² 안뜰은 옆으로는 공지가, 위로는 하늘이 보이도록 배치했다. 안뜰과 공지를 나누도록 설치한 수직 알루미늄 칸막이는 주변으로부터의 프라이버시 보호 및 보안 기능이 있다.

모든 방과 복도는 안뜰을 마주하고 있어 안뜰의 빛을 분산한다. 1층에는 주방, 화장실, 공용 공간이 있고, 2층에는 안뜰 양쪽으로 객실이 두 개 있다. 알루미늄 부등변 앵글 스틸 재질로 된 4m 수직 칸막이는 블레이드측은 안쪽을, 단측은 바깥쪽을 향하도록 배치했다. 이렇게 설계함으로써 방 안에서는 밖이 보이고 안뜰에서는 공지가 보이게 하는 반면, 밖에서는 안이 보이지 않도록 했다. 앵글 스틸을 빽빽이 배열하여 주변 환경이 스트립 모양으로 나뉘어 보인다. 게스트들은 주변 상황을 살펴볼 수 있고 문이 닫혀있어도 바깥과 연결돼 있는 느낌을 받을 수 있다.

몇 개의 선형 조명으로 분위기가 완전히 바뀌고 안뜰에 놓인 여러 개의 검은 돌에는 인상적인 그림자가 드리운다. 정원은 공지까지 수평으로 확장해 있어 실제 크기보다 더 넓어 보인다. 알루미늄 칸막이 상부와 2층 천장 들보 사이에는 어떠한 방해물도 없어 수직으로 하늘이 바로 올려다 보인다.

크고 작은 나무가 심어진 안뜰 바로 위로 조각 같은 하늘이 보인다. 게스트는 바람에 흔들리는 나뭇가지와 나뭇잎 사이로 보이는 하늘이 다양하게 바뀌는 모습을 만끽할 수 있다. 나무와 칸막이의 그림자가 하얀 실내에 선명하게 드리운다. 자연의 변화가 시간에 따라 담백하고 담담한 공간에 새겨진다. 게스트가 자신이 머무른 시간을 더 선명하게 기억할 수 있도록 빛과 바람이 가득한 공간을 만드는 것을 목표로 했다.

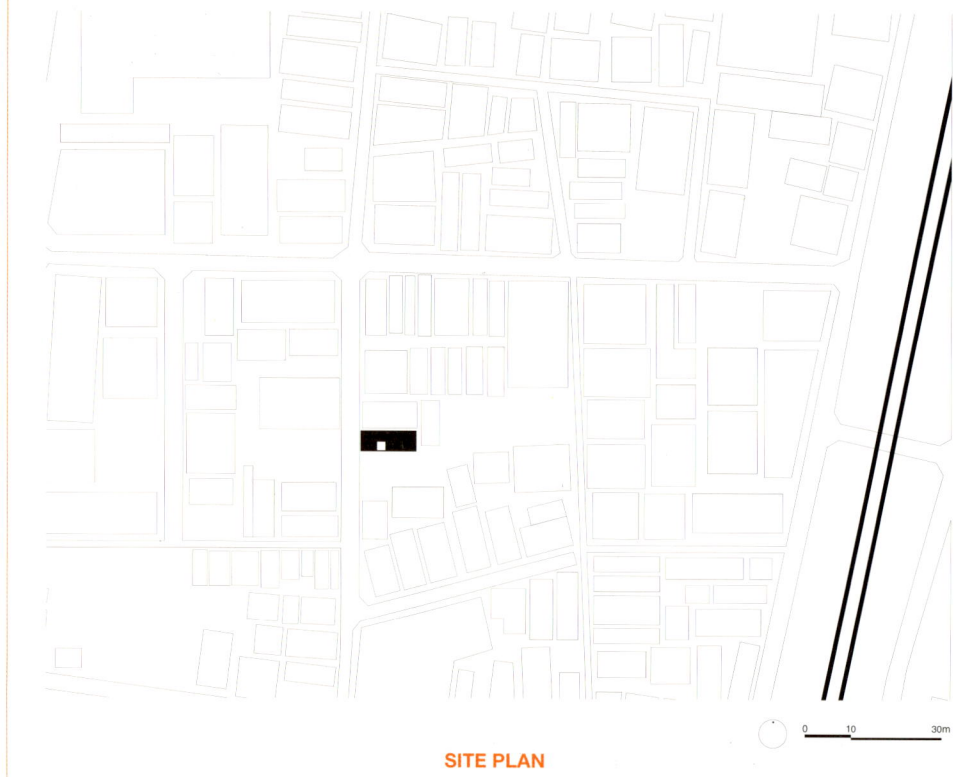

SITE PLAN

1 ROOM
2 PORCH
3 ENTRANCE
4 BATHROOM
5 DINING
6 KITCHEN
7 CORRIDOR

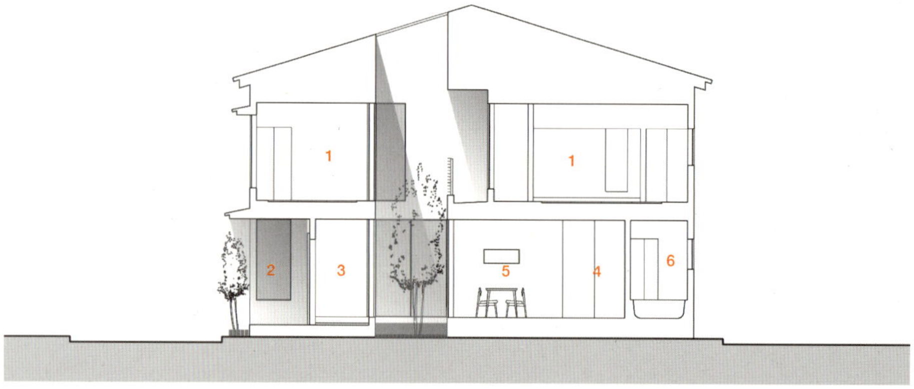

SECTION

Neighborhood Facility

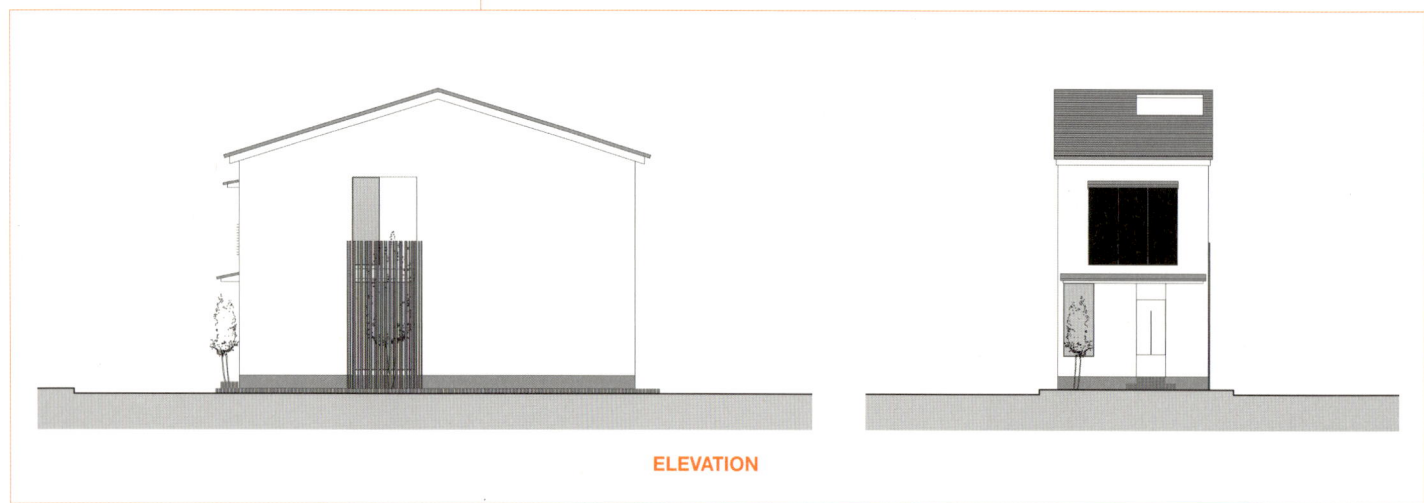

ELEVATION

Neighborhood Facility

DIAGRAM

1 HALL	6 LIVING ROOM	11 BATHROOM	16 CLOSET
2 PORCH	7 DINING ROOM	12 STORAGE	17 OPEN SPACE / PARKING
3 CORRIDOR	8 KITCHEN	13 ROOM	18 VOID
4 ENTRANCE	9 POWDER ROOM	14 SHOWER ROOM	
5 COURT	10 DRESSING ROOM	15 BALCONY	

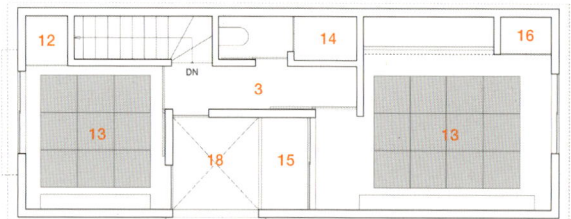

2ND FLOOR PLAN

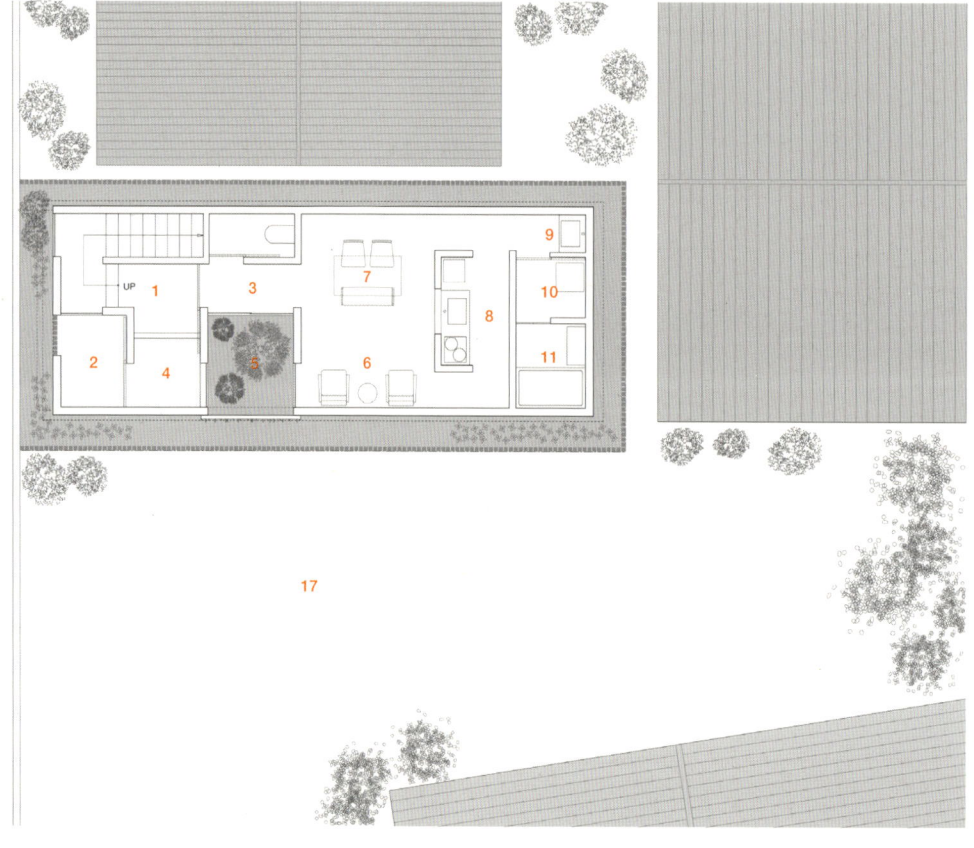

1ST FLOOR PLAN

Neighborhood Facility

Jun Murata / JAM

Born in Osaka, 1976 / Graduated Kindai University, 1999 / Worked at Takashi Yamaguchi & Associates / Founded JAM 2012 / Taught at several university / International Association of Designers / IAD, Member International Club for China Elite Designer / ICCED, International Advisor

Awarded Novum Design Award - Golden Prize _ France, 2020 / Minimalist Photography Awards - Honarable Mention _ Iran, 2020 / A' Design Award - Silver Prize _ Italy, 2020 / MUSE Creative Awards - Platinum Prize _ US, 2019 / Architecture Master Prize - Winner _ US, 2019 / DNA Paris Design Award - Honorable Mention _ France, 2019 / Architects of the Year "MINIMAL is MAXIMAL" - Winner _ Japan, 2018 / German Design Award - Special Mention _ Germany, 2016 / ICONIC Awards - Winner _ Germany, 2015 / Society of American Registered Architects Design Awards - Silver Prize _ US, 2014 / LICC | London International Creative Competition / Finalist _UK, 2014 / SBID International Design Excellence Awards - Finalist _ UK, 2014 / SPARK Awards - Gold Prize _ US, 2014

>> junmurata.com

Location
Ogin-dong, Jongno-gu, Seoul, Republic of Korea
Use
Bookstore, office
Site area
70.81m²
Building area
41.72m²
Total floor area
121.54m²

Floors
3F
Construction
Pod
Structure
R.C Rahmen
Exterior finish
Brick, STO
Interior finish
Exposed concrete, wood flooring

Design crews
Yim Sungwoo, Jo Minjung, Hur Sun
Consultants
Structure_Delta structure, MEP_Yousung
Photographer
Kim Changmook

Pi π 3.14

옥인동 79-2 근린생활시설
파이 3.14

어반아크
URBAN ARK

옥인동은 옥동에서 유래한 동명으로 옥류동, 혹은 옥골이라 불려지기도 하였는데, 지금의 옥인동과 통인동에 걸친 지역이다. 인왕산은 대대로 경치 좋은 명승지로 꼽혀왔고 그 자락에는 한양의 5대 명승지인 인왕동과 백운동이 있었다. 경복궁과 가까운 주택지이기도 해서 예부터 사람들이 많이 모여 살았고 양반과 중인들이 터를 물려가며 살았다. 청풍계 일대에는 양반들이, 옥류천 일대에는 중인들이 모였다. 경복궁 서쪽과 인왕산 사이의 일대의 서촌 지역은 중인문화의 중심공간이었다.
인왕산하면 가장 먼저 떠오르는 인왕제색도의 겸재 정선은 지금의 경복고등학교 자리인 청운동에서 태어나 화가로서 성공한 후에 옥인동으로 이사를 가서 살았다. 그가 비 갠 인왕산의 모습을

Neighborhood Facility

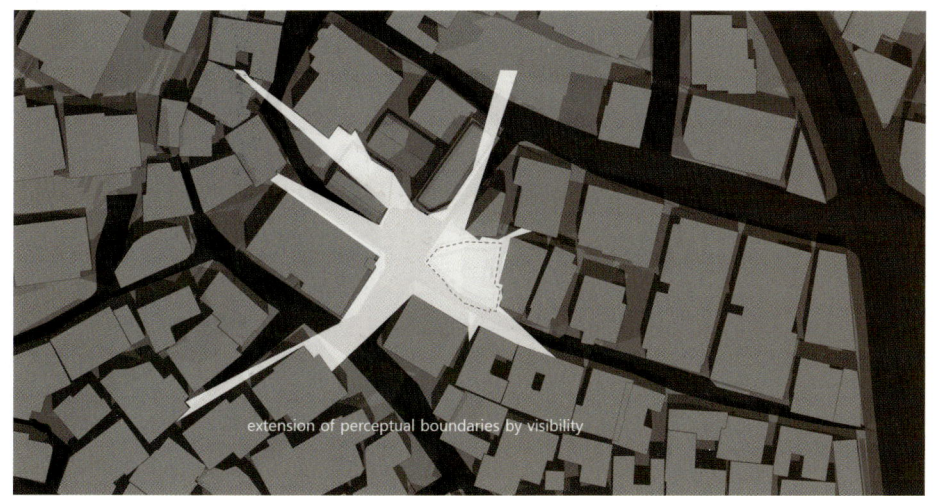

EXTENSION OF PERCEPTUAL BOUNDARIES BY VISIBILITY

MODEL

조감하여 인왕제색도를 그릴 수 있었던 것도 이곳을 잘 알고 있었기 때문에 가능했을 것이다. 또한 옥인동에 인곡정사라는 집을 짓고 살았는데, 늘 인왕산 주변과 골짜기를 소요하였다.

해당 대지는 서촌의 옥인동에 위치한 작은 쐐기모양의 땅이었다. 서촌의 구부러진 좁은 골목길들 사이 4개의 골목길이 만나는 곳에 형성된 작은 마을광장을 향하고 있다. 대지의 형태와 건축법에 의한 도로폭 확보, 모서리의 가각전제로 인하여 대지의 서쪽 모서리는 삼각형의 날 형태를 하게 된다.
서촌의 구불구불한 골목길 사이로 다닥다닥 붙어있는 오래된 건물들의 모습과 도시조직은 흥미로운 공간이다. 건물과 건물 사이의 골목을 공간의 주체로 바라본다면 골목은 또 다른 건축의 공간이 된다.
대상 부지를 중심으로 주변 골목은 우리 옛길의 스케일과 밀도를 유지하고 있다. 각 골목에서의 공간경험은 연속적이고 대상부지 앞의 작은 마을공터로 흐른다. 대상부지의 서쪽 모서리는 작은 마을공터를 향하고 있으며, 공터를 중심으로 4개의 골목길에서

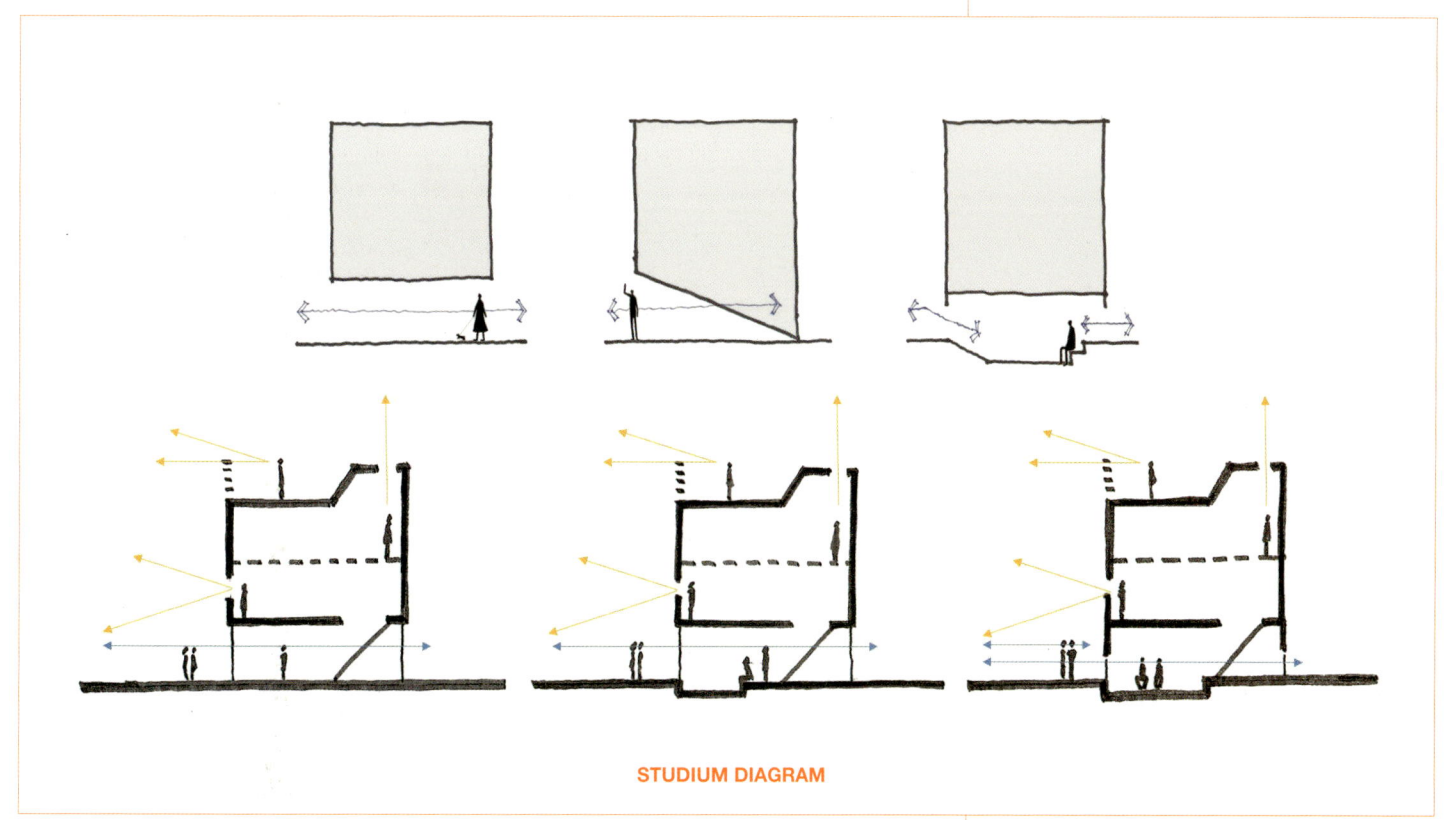

STUDIUM DIAGRAM

가장 먼저 마주하게 되는 영역이다. 대상부지의 경계를 확장하여 생각해본다면 오히려 외부로의 시각이 미치는 영역은 대상부지와 주변골목 및 건물들을 연결하게 된다. 우리는 대상지 주변 골목의 흐름을 건물 안으로 끌어들이고, 계단이라는 건축요소를 통해 골목의 공간경험을 수직적으로 연결, 확장시키는 전략을 구상하였다. 골목은 계단이 되고, 계단은 공간을 만들고, 그 공간이 건물이 된다. 프로젝트는 더 이상 근생이 아니라 골목의 연장이고 전망대가 되었다.

우리는 협소한 삼각형의 평면을 순환형으로 돌아가는 계단동선을 계획하였고, 최상층의 옥상에서 조망하게 될 인왕산과 남산의 모습을 고려하였다. 순환동선의 시퀀스에 있어서 보여지는 각 공간마다의 장면들을 세심하게 살폈고 공간요소에 맞게 띠창과 케이스먼트창을 적정하게 배치하였다. 최상층 지붕에는 꼬깔 모양의 천창을 두어 내부로 자연채광을 끌어들이는 동시에 옥상의 휴게테이블이 되도록 하였다.

The name of Ogin-dong originated from Ok-dong, also known as Okryu-dong or Okgol, which was the name of the area now covered by Ogin-dong and Tongin-dong. Inwangsan Mountain has been considered a scenic spot for generations, and at its foot lie Inwang-dong and Baekun-dong, which used to be among the five best scenic spots in Hanyang (ancient Seoul). Since it was also a residential area within close proximity to Gyeongbokgung Palace, many upper-class people had resided there for generations upon generations since ancient times. Nobles, called yangban, congregated in the Cheongpunggye district, and middle-class residents congregated in the Okryucheon district. The Seochon area between the west side of Gyeongbokgung Palace and Inwangsan was the epicenter of middle-class culture.

Gyeomjae Jeong Seon, best known for his

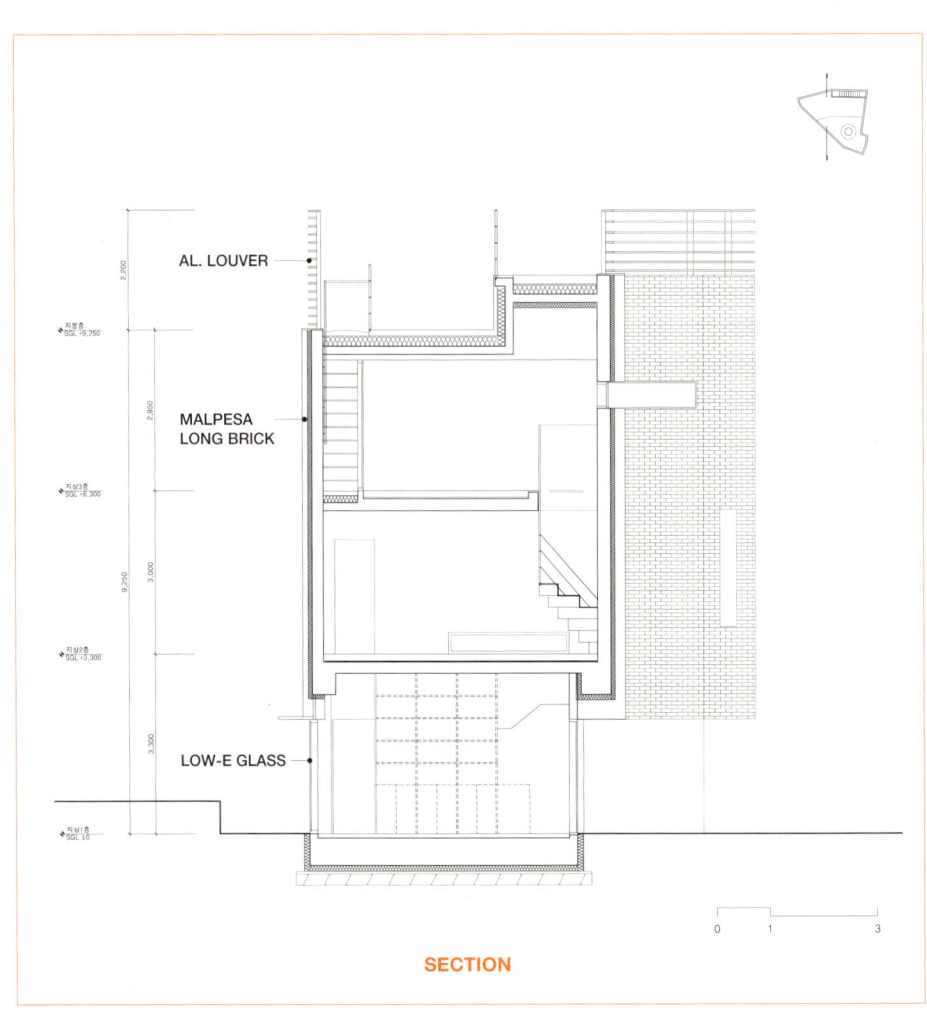

SECTION

Neighborhood Facility

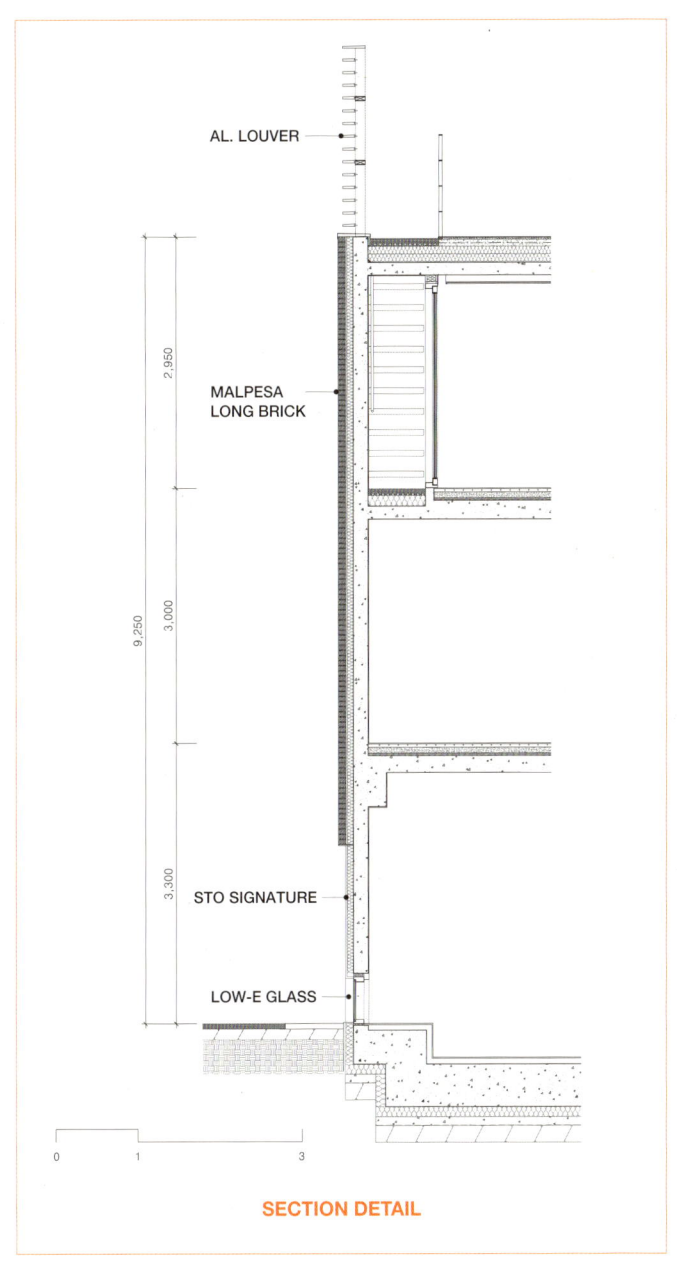

SECTION DETAIL

- AL. LOUVER
- MALPESA LONG BRICK
- STO SIGNATURE
- LOW-E GLASS

painting Inwangjesaekdo, is the first person who springs to mind when it comes to Inwangsan Mountain. He was born at the present location of Gyeongbok High School in Cheongun-dong and later moved to Ogin-dong as a successful painter. He was able to paint Inwangjesaekdo from a bird's-eye view of Inwangsan after the rain probably because he knew this area like the palm of his hand. He lived in a house called Ingok Jeongsa in Ogin-dong, and he often rambled around Inwangsan and its valleys.

The location was a small, wedge-shaped plot of land located in Ogin-dong, Seochon. It faces a tiny town square formed by four alleyways that converge amid Seochon's many narrow, curved alleyways. The western corner of the site took the shape of a triangular edge due to the shape of the site, the securing of the road width required by the building law, and the clearance at the angle of the corner.

The appearance and urban fabric of the old buildings closely joined to one another in Seochon's winding alleys make it a fascinating place. When we consider the alleyways amid buildings as spatial subjects, they become architectural spaces as well.

Neighborhood Facility

1 BOOKSTORE
2 COUNTER
3 DESIGN OFFICE
4 STORAGE
5 RESTING SPACE
6 OBSERVATION ROOFTOP

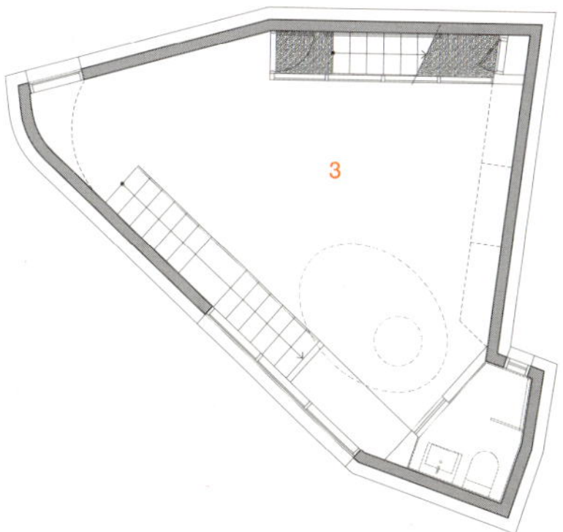

3RD FLOOR PLAN

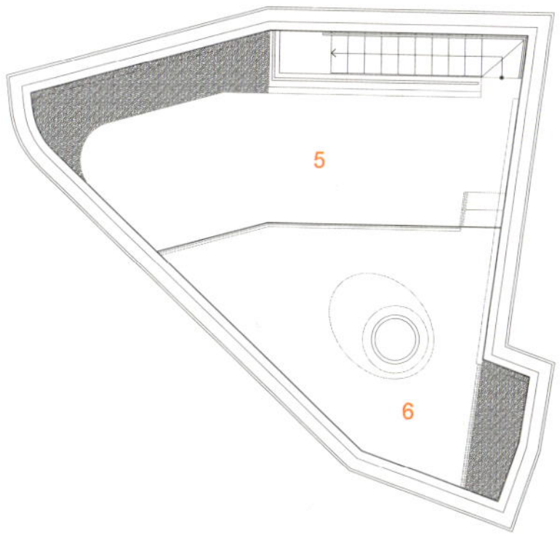

ROOF PLAN

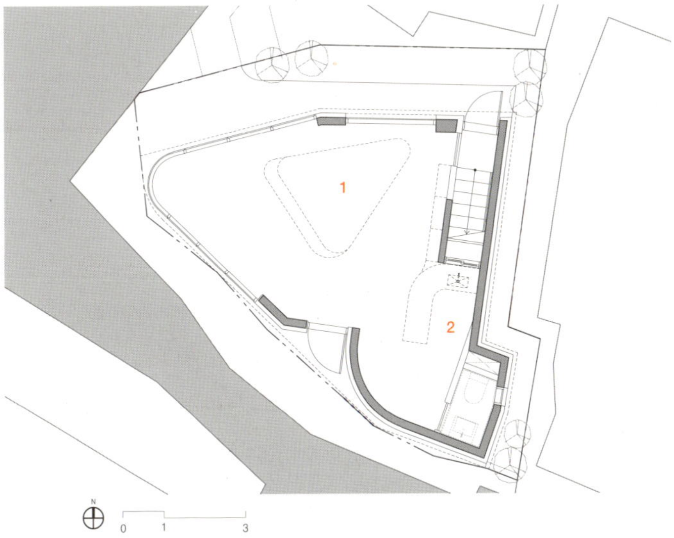

1ST FLOOR PLAN

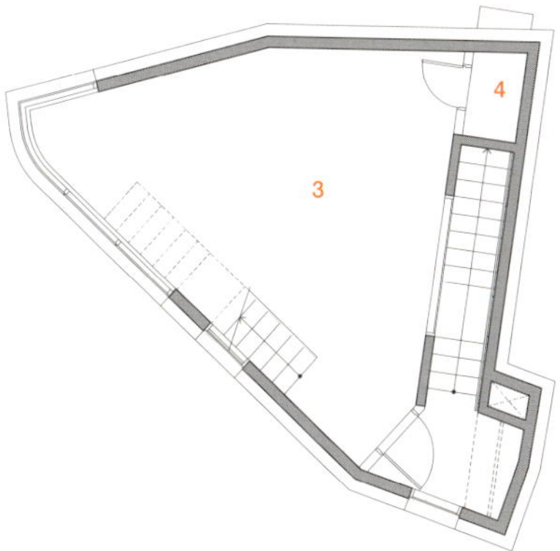

2ND FLOOR PLAN

The alleys around the target location preserve the scale and density of our historic roads. The spatial experience in each alley is continuous and flows into the little town square in front of the target location. The western corner of the target location faces the little town square, and it is the first spot encountered in the four alleyways stretching from the square. Considering the expansion of the target location's boundaries, the visible exterior areas will connect it with the surrounding alleys and buildings. We came up with a strategy to invite the flow of alleyways around the location into the building and vertically connect and elevate the spatial experience of the alleys by utilizing the stairs as an architectural element. Alleys become stairways, the stairways create spaces, and the spaces become a building. The project is no longer limited to a certain area but rather becomes an extension of the alleyways and then an observatory.

We designed a circular traffic flow through stairways on a narrow triangular plane while taking into account the rooftop views of the Inwangsan and Namsan mountains. The views from each space in the circular traffic sequence were carefully examined, and the strip and casement windows were properly aligned with the spatial elements. A cone-shaped skylight was installed on the rooftop to bring natural light inside while also serving as a resting bench.

Neighborhood Facility

urban ark
design navigators for a better world

어반아크
urban ark

"평범함 속에 잠재된 가치의 추구"
어반아크는 공간의 경험을 조직하여, 가치 있는 장소로 만드는 디자인 회사이다.
피상적 건축 담론이나 자본에 종속된 개발의 논리가 아닌, 우리의 평범한 일상 속에 숨어있는 잠재된 가치를 드러내어 디자인에 반영하는 작업을 추구한다.
어반아크의 디자인 과정은 도로를 달리는 자동차 운전보다는 파도를 타는 비선형의 윈드서핑과 같다. 그 과정은 풍향에 따라 수시로 몸을 움직이고 돛의 방향을 틀어야 하는 불확정적인 이동의 경로이다. 이것은 의도하지 않은 우연의 연속이며, 동시에 앞으로 나아갈 수 있게 하는 동력이 됩니다. 마찬가지로 프로젝트를 통해 요구되는 기본적인 조건 외, 프로젝트 이면의 일상에서 발견되는 우연의 단서들은 어반아크의 상상력과 만나 고유의 가치를 갖는 공간으로 진화한다. I.M.A.G.E.는 이러한 디자인 항해를 가능케 하는 어반아크의 문제 해결 프로세스다.

"Discovering hidden dimension in everyday life"
'urban ark' is a design practice based in Seoul, South Korea that organizes the experience of space and makes it a valuable place. Rather than pursuing superficial architectural discourse or capital-dependent development logic, they seek to reveal the potential values hidden in their ordinary daily lives and reflect them in the design.
Their design process is more like non-linear windsurfing on the waves than driving a car on the fixed road. The process is an indeterminate path of movement that requires frequent body balancing and change the direction of the sail according to the wind direction. This is a series of unintended coincidences, and at the same time, it generates the driving force that keeps us moving forward. Similarly, in addition to the basic conditions required through the project, the clues of chance found in the daily life behind the project meet our imagination and evolve into a space with irreplaceable values. I.M.A.G.E. is urban ark's problem-solving process that makes this design voyage possible.

Location
Saitama Prefecture, Japan
Use
Bar
Site area
200m²
Built area
42m²
Total floor area
42m²

Floors
1F
Exterior finish
Over-painting of existing wall surface
Interior finish
Larch plywood, oiled
Project architect
Yoshinori Hasegawa, Fumie Horikoshi
Photographer
Naomichi Sode

Restaurant in Ogawa

오가와 레스토랑

UENOA ARCHITECTS

Neighborhood Facility

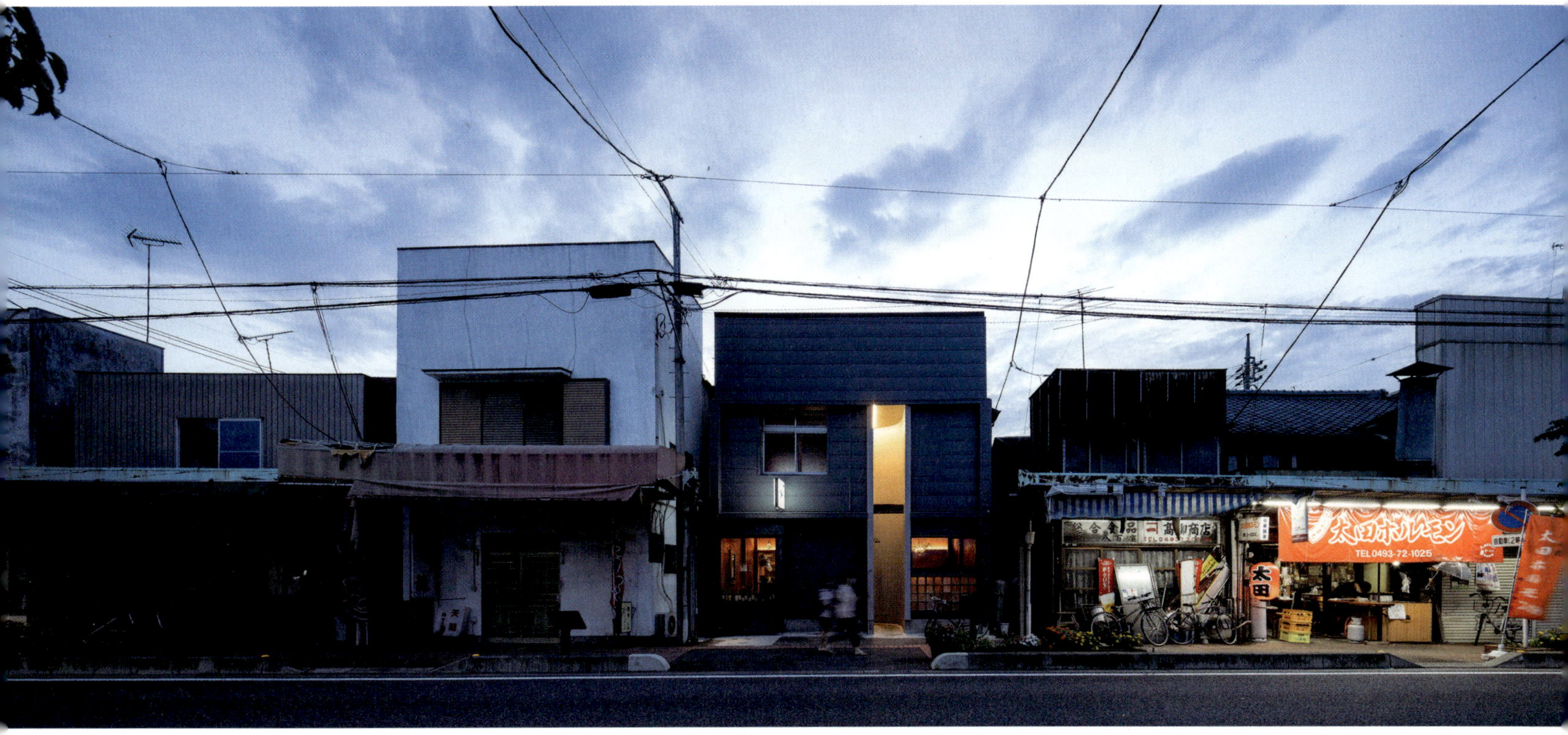

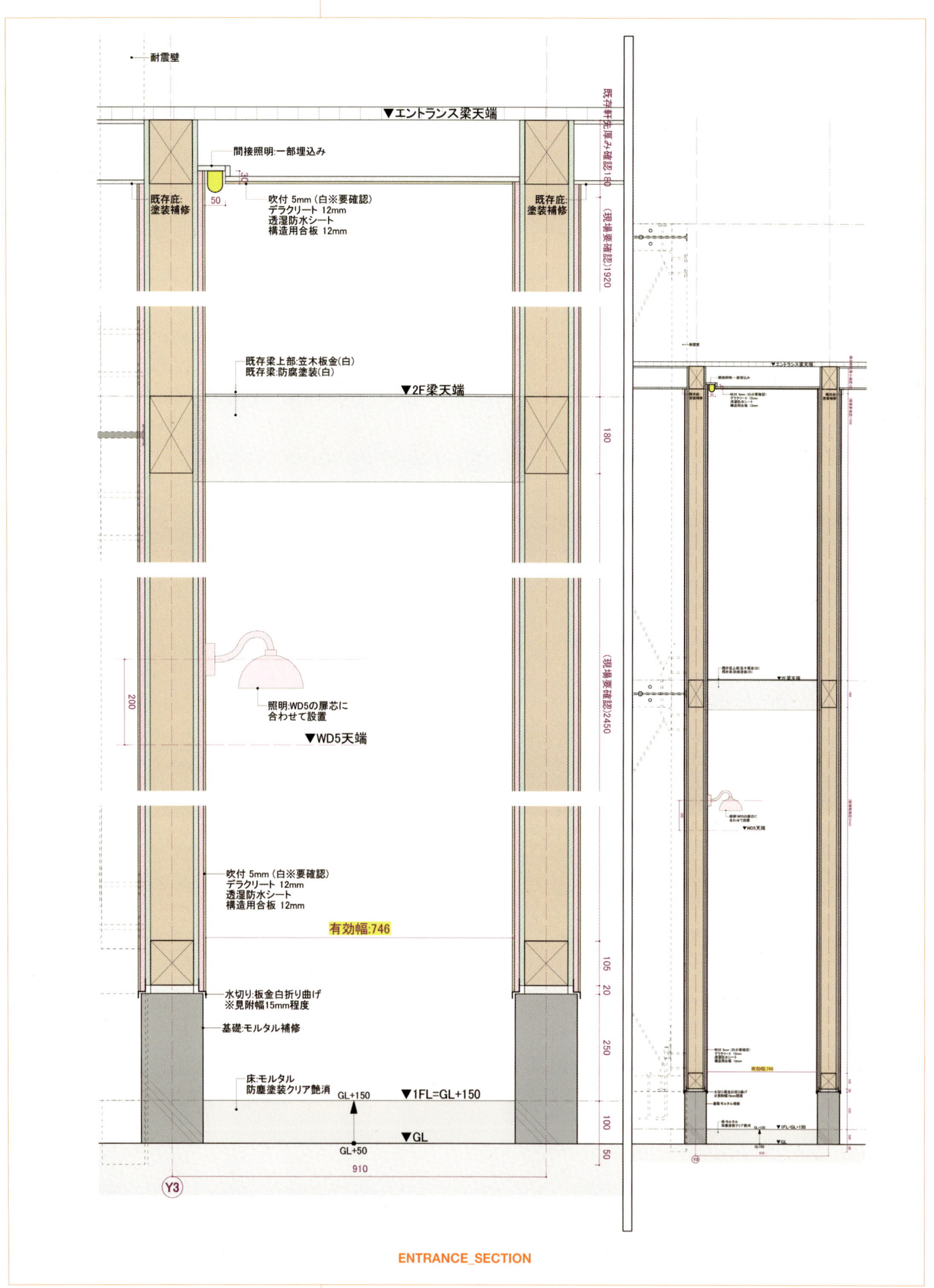

ENTRANCE_SECTION

Neighborhood Facility

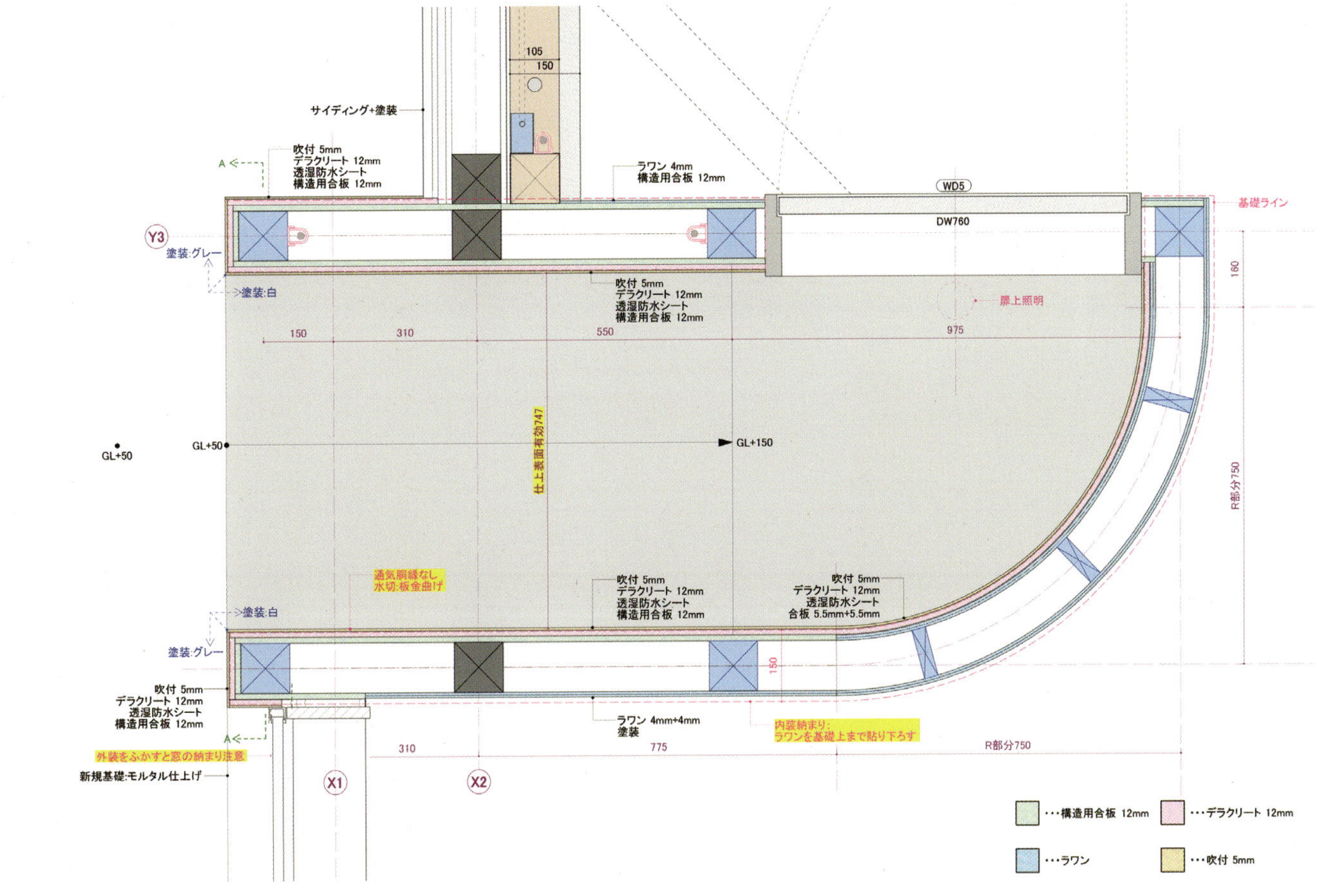

ENTRANCE_DETAIL

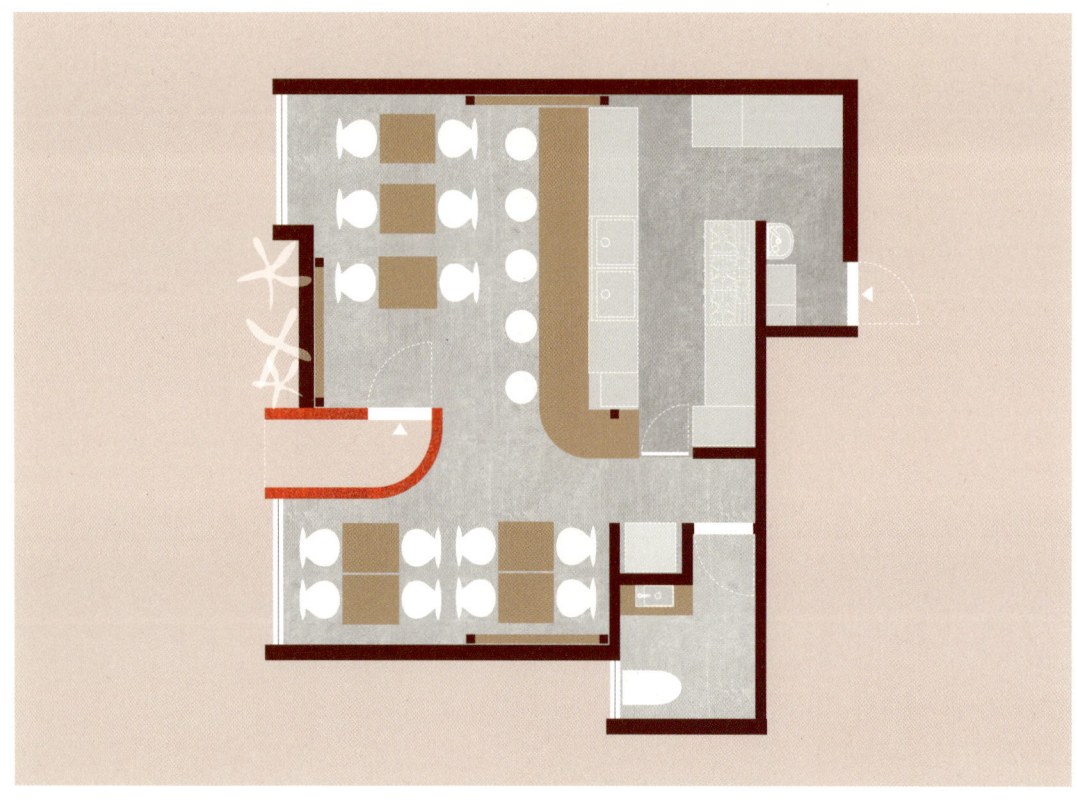

PLAN

Neighborhood Facility

Creating a new face for the shopping district

This project involves the renovation of an 80 year old house with a store in a corner of a shopping street in Ogawa-machi, Saitama Prefecture. The client's grandmother used to run a beauty salon here, and the client, who used to run an Italian restaurant in Tokyo, wanted to move there and open a restaurant. At the same time as the project began, it was decided to phase out the arcade over the sidewalk in the shopping arcade. The arcade was built at the same height as one level of the building, and the two-story buildings lining the shopping street were divided into two halves by the arcade, but the removal of the arcade revealed the height of each building for the first time. We wondered if we could use this as a trigger for our planning, and decided to create an entrance as an external space with the height of two stories, which had been difficult to experience before. The entrance space, which was created by inserting two layers of blank space from the outside, creates an impressive face of the building and at the same time functions as a rain shelter instead of an arcade, and also serves as a structural reinforcement for the existing building. The interior space is gently divided by the inserted entrance space, creating a cozy atmosphere that is visible and hidden from the restaurant's seating area.

상가에 새로운 얼굴을 만들다

본 프로젝트는 사이타마현 오가와 마치 상가 거리 구석에 자리잡은 80년된 겸용 주택을 개조하는 작업이다.

이곳은 건축주의 할머니가 과거 미용실을 운영했던 곳으로, 도쿄에서 이태리 레스토랑을 운영하던 건축주는 이곳으로 옮겨와 식당을 열고 싶어했다.

본 프로젝트의 시작과 동시에 상가 보도 위에 있는 아케이드를 단계적으로 철거하기로 결정했다.

아케이드는 빌딩 1층 높이였고, 쇼핑 거리에 늘어선 2층 건물은 아케이드로 인해 둘로 나뉘어 있었는데 아케이드를 철거하자 건물은 처음으로 그 높이를 드러냈다.

이를 기획의 계기로 삼을 수 있을지 고민한 끝에 출입구를 전에는 보기 힘들었던 2층 높이의 외부 공간으로 설계하기로 결정했다. 바깥쪽에서 두 층의 여백을 추가해 마련한 출입구 공간은 건물의 얼굴을 인상적으로 만드는 동시에 아케이드를 대신해 비막이 역할과 기존 건물의 구조 보강 역할을 한다. 내부 공간은 출입구 공간의 추가로 완만하게 분리되어 아늑한 분위기를 자아내고 식당 테이블에 앉으면 출입구가 보이지 않도록 계획했다.

Neighborhood Facility

Yoshinori Hasegawa + Fumie Horikoshi / UENOA architects

UENOA is a design office represented by Yoshinori Hasegawa and Fumie Horikoshi. We are doing architectural design on the theme of "re-designing living".
In the 20th century, mankind has realized that the modernization of architecture can enrich lives. We believe that what is important for 21st century architecture is how to achieve a sustainable environment and how to have "uniqueness"
So, we feel the potential of wooden construction using new technology. Wooden construction has the potential to bring circulation to the declining forests and to rediscover the local vernacular technologies.
>> uenoa.com

Location
Seongdong-gu, Seoul, Republic of Korea
Use
Commercial
Site area
80.57m²
Built area
45.92m²
Total floor area
202.19m²
Floors
B1F, 4F

Exterior finish
Exterior insulation finish system
Interior finish
Paint
Project architect
Lee Keun Sik
Construction
P·S construction engineering
Photographer
Gu Uijin

JUNG bldg.

정빌딩

엘케이에스에이 건축사사무소
LKSA

성수동 준공업지역 골목길에 있는 대지이며, 90년도 지어진 연와조 구조의 지하1층, 지상1,2층의 주거 건물을 근생 건물로 새 생명을 부여하는 프로젝트였다. 주거 용도의 건물이기에 용도변경을 실행하여 근생 용도로 전환한 다음, 주거 양태의 공간 구성을 근생 용도에 부합하도록 대공간으로 재구성하며, 2개층 증축을 더하여 진행하는 복합 허가 프로젝트였다.

기존건물의 구조가 연와조이기때문에 대공간으로 구성하기 위해 철골 구조를 주 구조로 치환하였다. 천장고의 최대 확보를 위해 철골의 사이즈 계획을 최소화하였다. 또한 기존 지하층의 기초도 매우 부실한 상태여서 기초 보강을 위해 증타를 계획하고 시공하였다.

건물의 입면도 통유리로 계획하여 최대한 근생의 가치를 살릴 수 있도록 구성하였다. 주차 추가 없는 2개층의 증축 구성을 위하여 면적 조정을 하면서 3층과 4층에 건축의 외부공간인 테라스를 구성하여 높은 레벨에서의 성수동의 도시적 분위기를 구가할 수 있도록 디자인하였다. 성수동이 붉은 벽돌로 구성된 건물이 많기때문에 그 가운데에서 본 근생건물만의 시지각적 인지성을 고양시키기 위해서 블랙의 외단열 시스템을 이용하여 솔리드 부분을 계획하였다.

오랜 세월동안 주거의 양태로 존재해왔지만, 새로운 프로그램(용도)과 그에 걸맞는 내외부 공간구성을 통하여, 급변하는 도시 안에서 새로운 숨결을 간직한 채, 역동적이고, 활력있는 역할을 수행할 수 있기를 바래본다. 그리하여 도시의 생명력 넘치는 매력있는 인자로 작용할 수 있기를 기도해본다.

BEFORE

DISMANTLING BUILDING 해체작업

Neighborhood Facility

STRUCTURAL REINFORCING 구조보강

The location is in an alleyway in Seongsu-dong's semi-industrial district. The goal of the project was to breathe new life into a residential brick building built in the 1990s with a basement and two stories above ground. Since the building was for residential use, the project had gone through a complex permit procedure in which its use was modified to community use, and then the residential space composition was reconstructed into a larger space for community usage with the addition of two more stories.

Since the original structure was brick, the primary structure had been rebuilt with a steel frame to provide greater space. The size of the steel frame was minimized to achieve the maximum ceiling height. In addition, since the foundation of the existing basement floor was in such bad condition, additional pilings were planned and installed to reinforce the foundation.

The building's façade was designed with glass to emphasize its value as a neighborhood facility. The third and fourth floors were planned to have the urban feel of Seongsu-dong through the construction of external spaces like terraces while adjusting the area for the structure with two more stories without extra parking. Since there are numerous buildings made of red bricks in Seongsu-dong, solid parts were designed to employ a black external insulation system to enhance the visual perception of this neighborhood living facility among the surrounding buildings.

Despite the fact that it has been a residential place for a long time, we hope that it will play a dynamic and energetic role through a new program (purpose) and internal and external spatial structures appropriate for it, while providing a breath of fresh air and acting as an attractive factor brimming with vitality in the rapidly changing city.

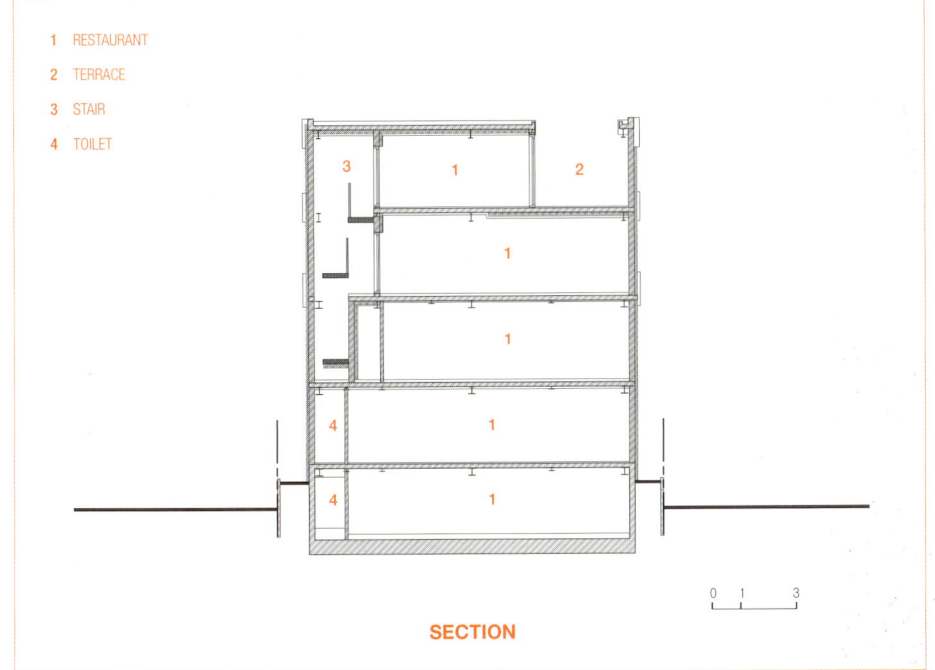

1. RESTAURANT
2. TERRACE
3. STAIR
4. TOILET

SECTION

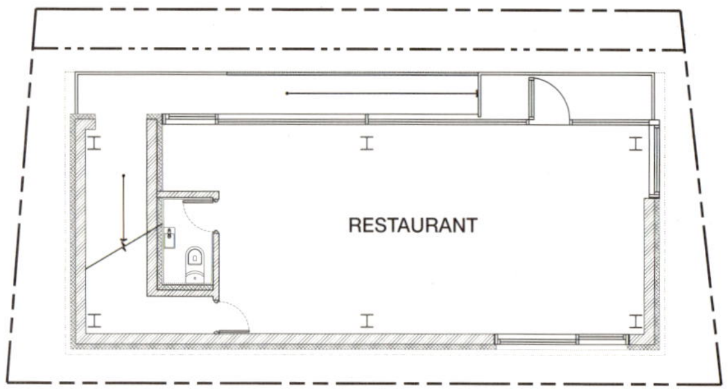

2ND FLOOR PLAN

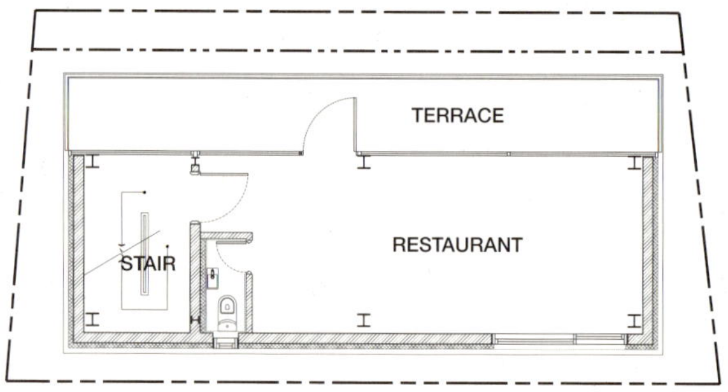

3RD FLOOR PLAN

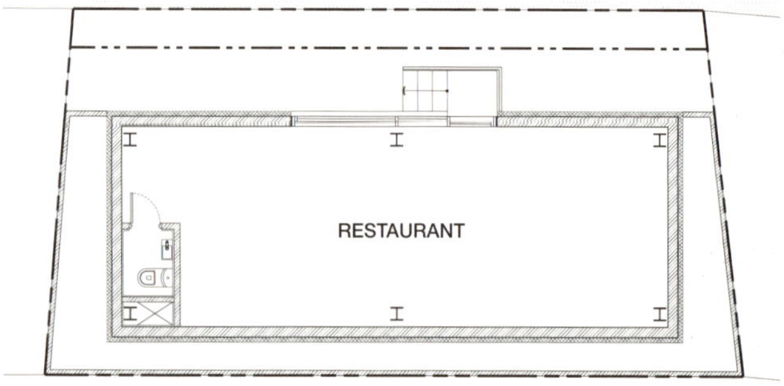

B1 FLOOR PLAN

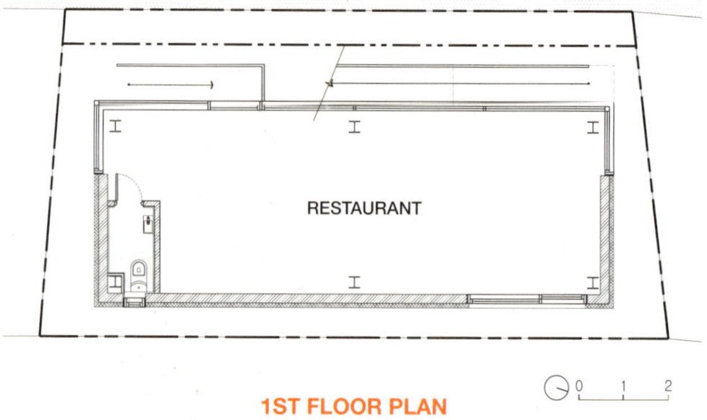

1ST FLOOR PLAN

1 RESTAURANT
2 TERRACE

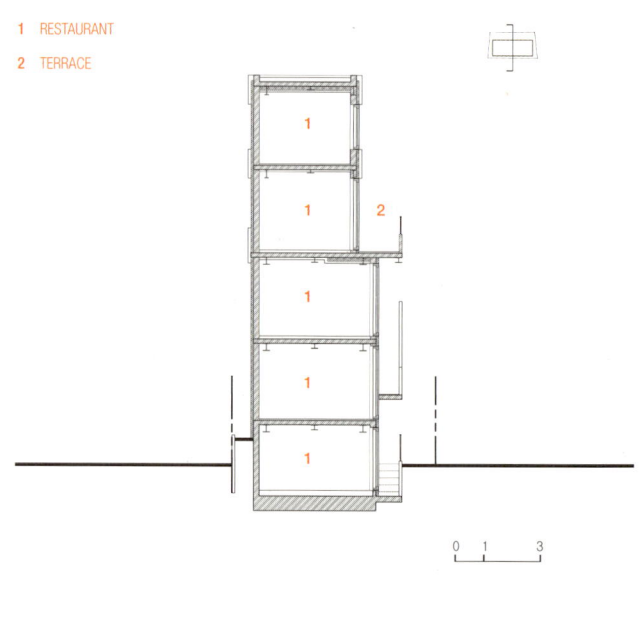

SECTION

Neighborhood Facility

이근식 / LKSA 건축사사무소
Lee Keunsik / Lee Keun Sik Architects

LKSA 건축사사무소는 '건축가의 본질적인 의무와 책임은 건축주를 비롯한 그 공간을 향유하는 사람들의 행복을 위한 노력에 있다'라는 신념 하에 2012년부터 매 순간 건축을 향한 깊은 애정과 장인정신 그리고 소명 의식을 갖고 설계한다. 건축뿐만 아니라 인테리어, 가구, 조경, 사업 컨설팅까지 건축에 관여하는 모든 요소의 전문성을 가지고 있다. 이 모든 요소가 건축가의 일관된 사고에서 연속성을 가질 때 비로소 삶을 위한 그릇이 현실화된다고 믿는다.
이근식 건축가는 한양대학교를 건축학과를 졸업하고 삼우종합건축사사무소에서 실무를 쌓았으며, 2012년부터 LKSA 건축사사무소를 운영해오고 있다. 2020년에 '대한건축사협회 신진건축사상'을 수상했다.
>> www.lksa.kr

NIKI BAKERY

니키 베이커리

KAMAKURA STUDIO

Location
Ueno, Tokyo
Use
Bakery
Site area
55m²
Built area
50m²
Total floor area
102m²
Floors
B2 , 2F

Exterior finish
Alpolic
Interior finish
Tile, Melamine veneer
Project architect
Keisuke Fukui, Keisuke Morikawa
Photographer
Keisuke Fukui, Keisuke Morikawa

We thought about the communication between the people of the city and the architecture. One of the most important requests from the client was to use the panda, a symbol of the vitality of the city of Ueno, as a design motif. They are a bakery, but they wanted to emphasize the panda. We responded to this contradictory request by coming up with a facade design that looks like a panda from a distance, but when approached, the panda's face is composed of a small bread sign. Panda and bread are very similar in Japanese pronunciation. Bread is pronounced "pan". In Japanese, "da" is often added to the end of a word when pointing to something. In other words, when you point to bread, you pronounce it "pan-da". It is the same pronunciation as "panda". Many tourists and locals see this bakery and notice the "panda" when they look at the bakery from a distance. And when they approach, they notice the little bread and say "pan-da!". This proposal is to create such smiling communication between the people of the city and the architecture.

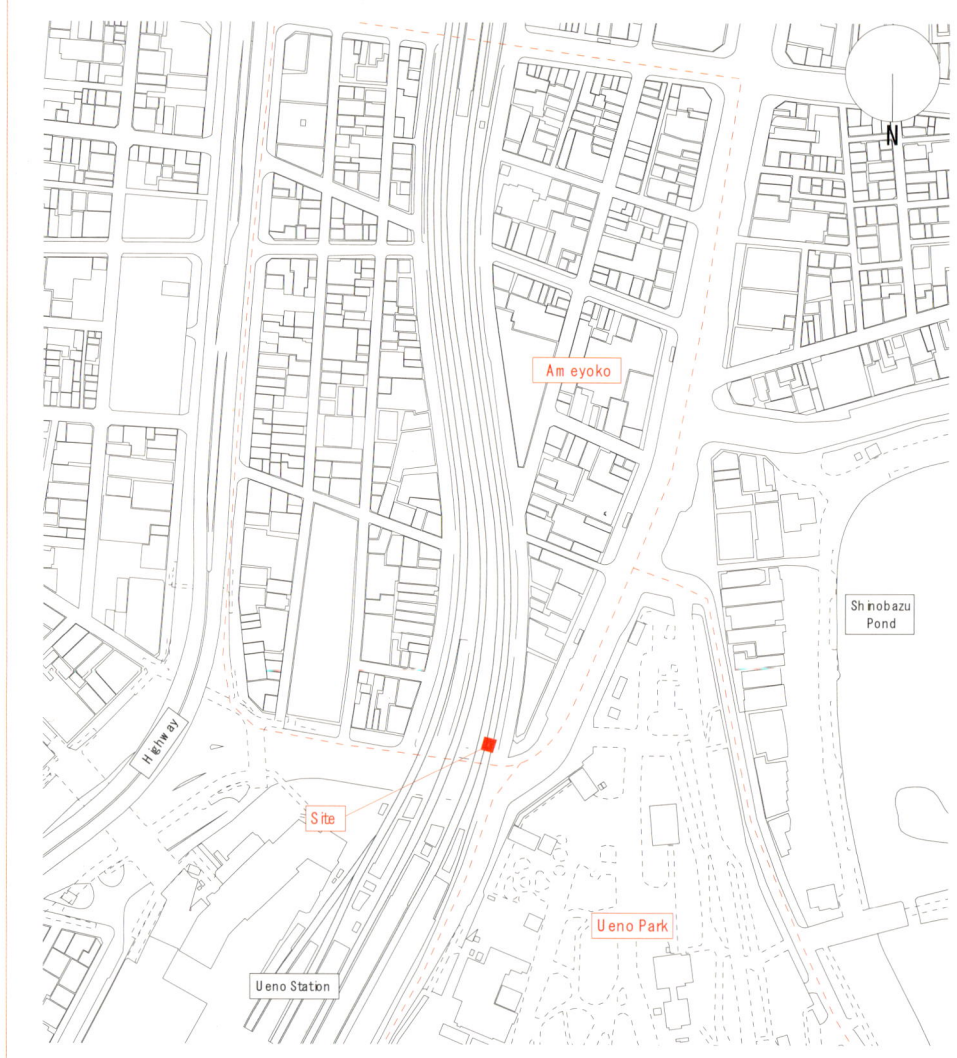

SITE PLAN

해당 대지는 도쿄 우에노역 바로 앞 고가 철로 아래에 위치해 있다. 아메요코와 우에노 공원의 경계에 위치한다는 점에서 독특하다. 아메요코는 제2차 세계대전 이후 암시장으로 시작하여 다양한 상품과 수많은 다국적 상점들이 남아있다. 530,000m2 규모의 우에노 공원은 6개의 박물관, 판다가 있는 동물원, 도쿄 예술 대학 등 문화 시설이 밀집해 있다. 그 결과 아메요코와 우에노 공원 주변 지역은 연간 2,000만 명 이상의 관광객을 끌어들이며 수많은 사람들이 이곳을 지난다. 프로젝트는 현장에 있는 기존 카페를 개조하여 베이커리를 만드는 것이었다.

우에노에는 수많은 판다가 살고 있다. 동물원 뿐만 아니라 도시에서도 조각상, 우편함 등으로 존재하며 도시와 조화를 이루고 있다. 이 독특한 우에노 컨텍스트를 파사드 디자인에 담아, 빵집 간판 중 몇 개에 팬더 사인을 적용하였다.

The site is located right in front of Ueno Station in Tokyo, under the elevated railway tracks. It is unique in that it is located on the boundary between Ameyoko and Ueno Park. Ameyoko began as a black market after World War II and remains a diverse assortment of goods and a large number of multi-national stores. Ueno Park is a 530,000 m2 site with a concentration of cultural facilities, including six museums, a zoo with pandas, and the Tokyo University of the Arts. As a result, the area around Ameyoko and Ueno Park attracts more than 20 million tourists annually, and numerous people pass through the site as well. The project was to renovate an existing café on the site and create a bakery. Ueno is home to numerous pandas. They are not only in the zoo, but also in the city as statues, stuffed animals, and mailboxes. Pandas blend into the city. We incorporated this unique Ueno context into the façade design, mixing only a few panda signs among the numerous bread signs.

Neighborhood Facility

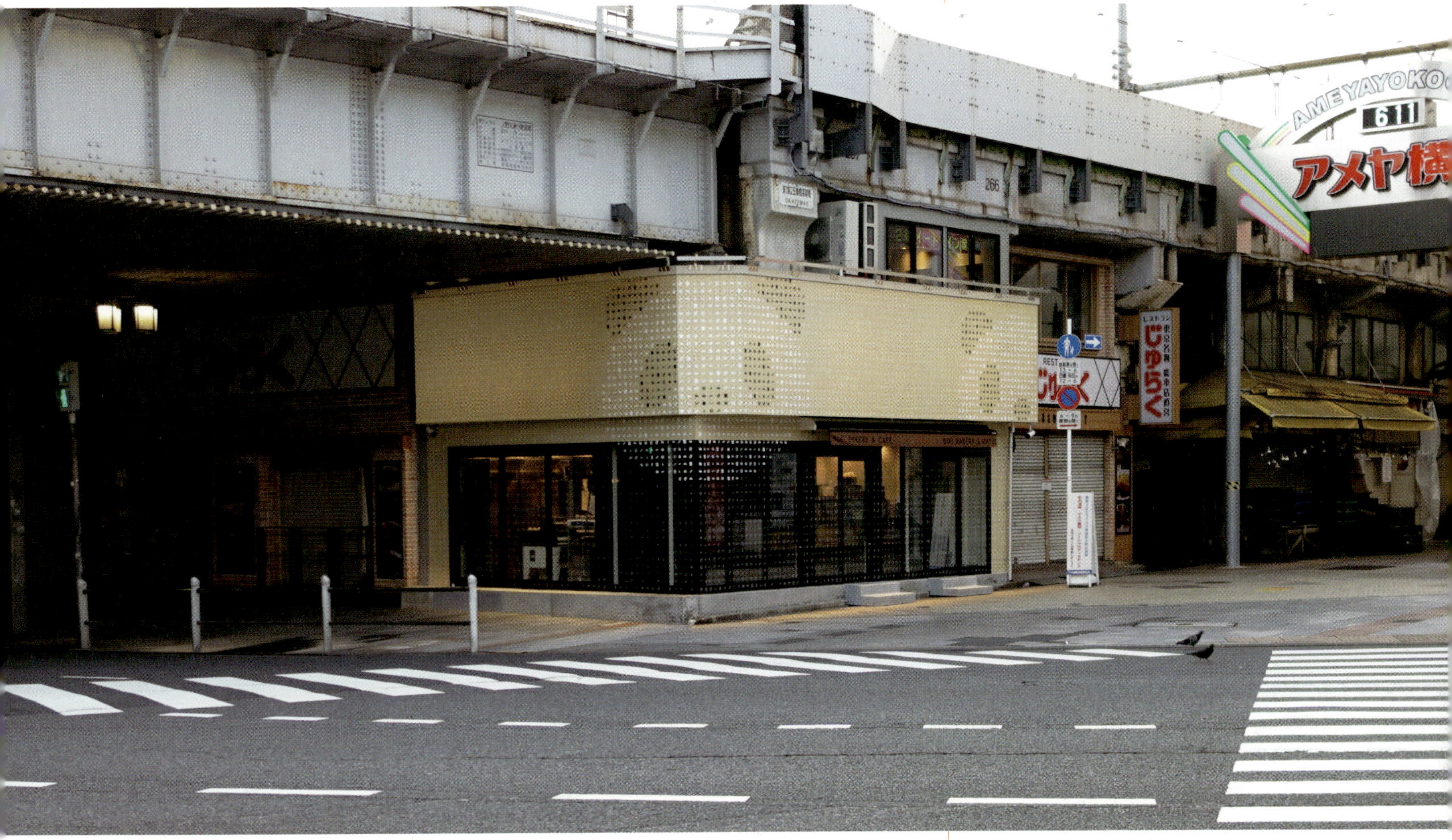
AMEYAYOKO
611
アメヤ横
じゅらく

AMEYAYOKOCHO
アメヤ横丁
マスク着用で楽しいお買物!!
RESTAURANT
じゅらく
PRONTO
PRONTO
じゅらく
スギドラッ

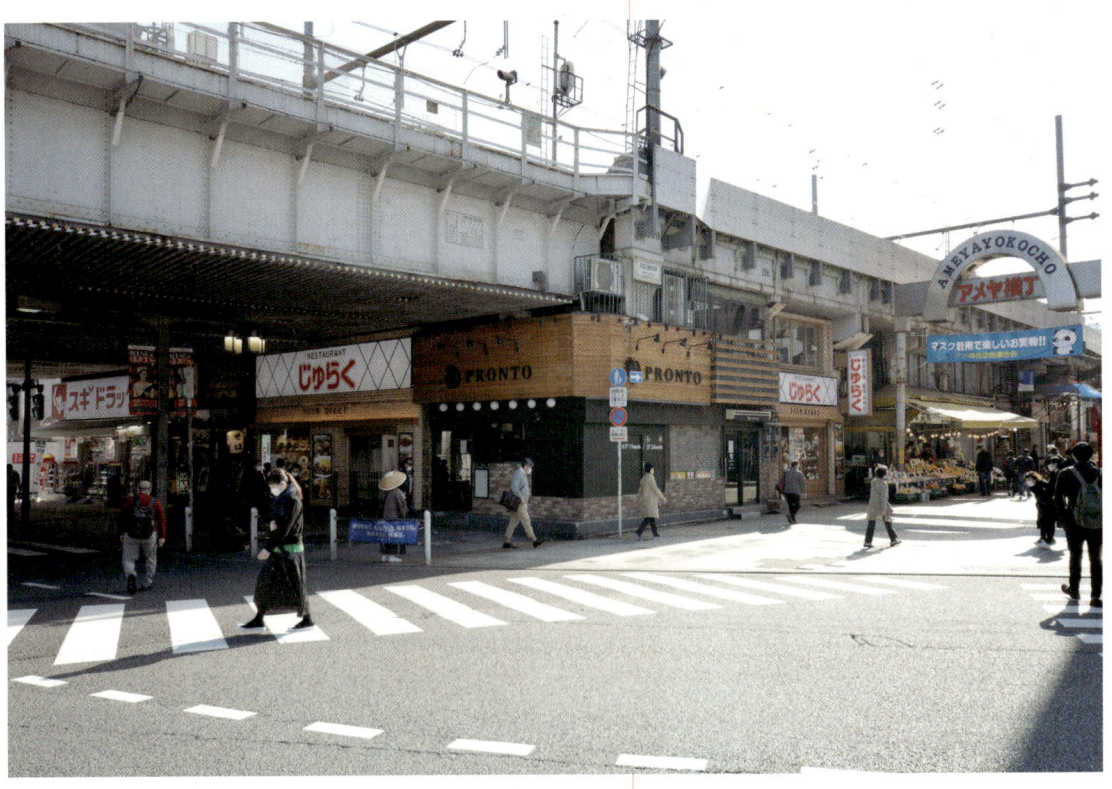

BEFORE

이 프로젝트에서 우리는 도시 사람들과 건축의 소통에 대해 생각했다. 고객의 가장 중요한 요구사항은 우에노시의 활력을 상징하는 판다를 디자인 모티브로 사용하는 것이었다. 베이커리였지만 판다를 강조하고 싶어했다. 우리는 이 상반되는 요구사항을 멀리서 보면 판다처럼 보이고 가까이 다가가면 판다 얼굴이 작은 빵 모양이 되는 파사드 디자인으로 풀어나갔다. 판다와 빵의 일본어 발음은 매우 비슷하다. 빵은 "팬"이라고 발음한다. 일본어에서 "다"는 무언가를 가리키는 말의 어미로 쓰인다. 즉, 빵을 가리킬 때 "판-다"라고 발음하게 된다. "판다"와 발음이 같다. 멀리서 이 빵집을 보면 "판다"처럼 보인다. 그래서 다가가보면 작은 빵인 것을 알아차리고 "판-다"라고 말한다. 이 제안의 목적은 도시 사람들과 건축 간에 미소 짓는 소통을 만드는 것이다.

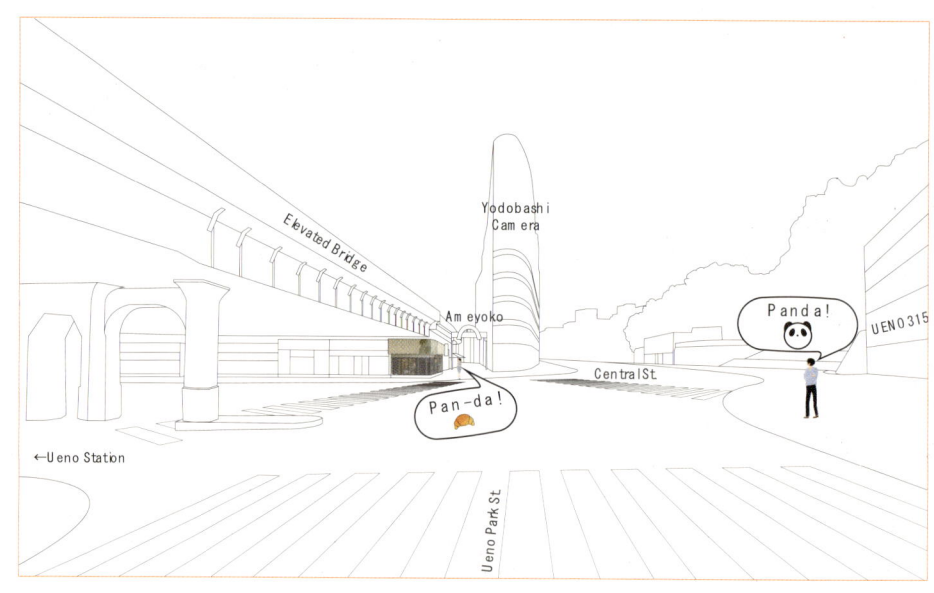

PERSPECTIVE

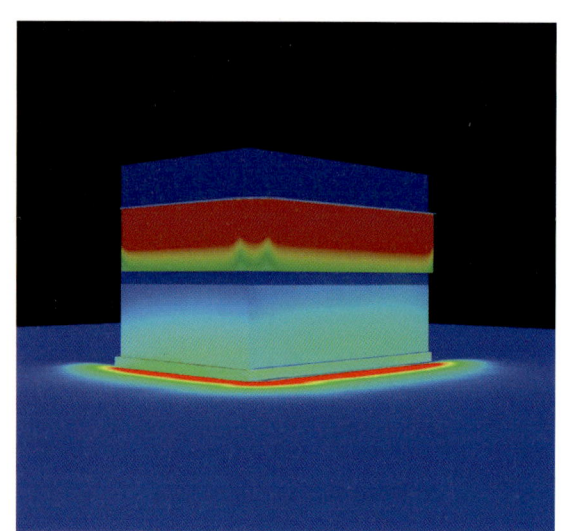

LIGHTING TO THE STREET

조명의 각도는 파사드의 간판뿐만 아니라 사람들이 끊임없이 지나다니는 빵집 앞 거리를 비추도록 조절했다. 점등계획은 파사드 전체가 균일하게 밝아지도록 설계하였으며, 편의점처럼 너무 밝지 않게 하였다.

The angle of the lighting was adjusted to not only illuminate the sign on the facade, but also the street in front of the bakery, where people constantly pass by. The lighting plan is designed so that the entire facade is evenly and uniformly brightened, and not too bright like a convenience store.

Neighborhood Facility

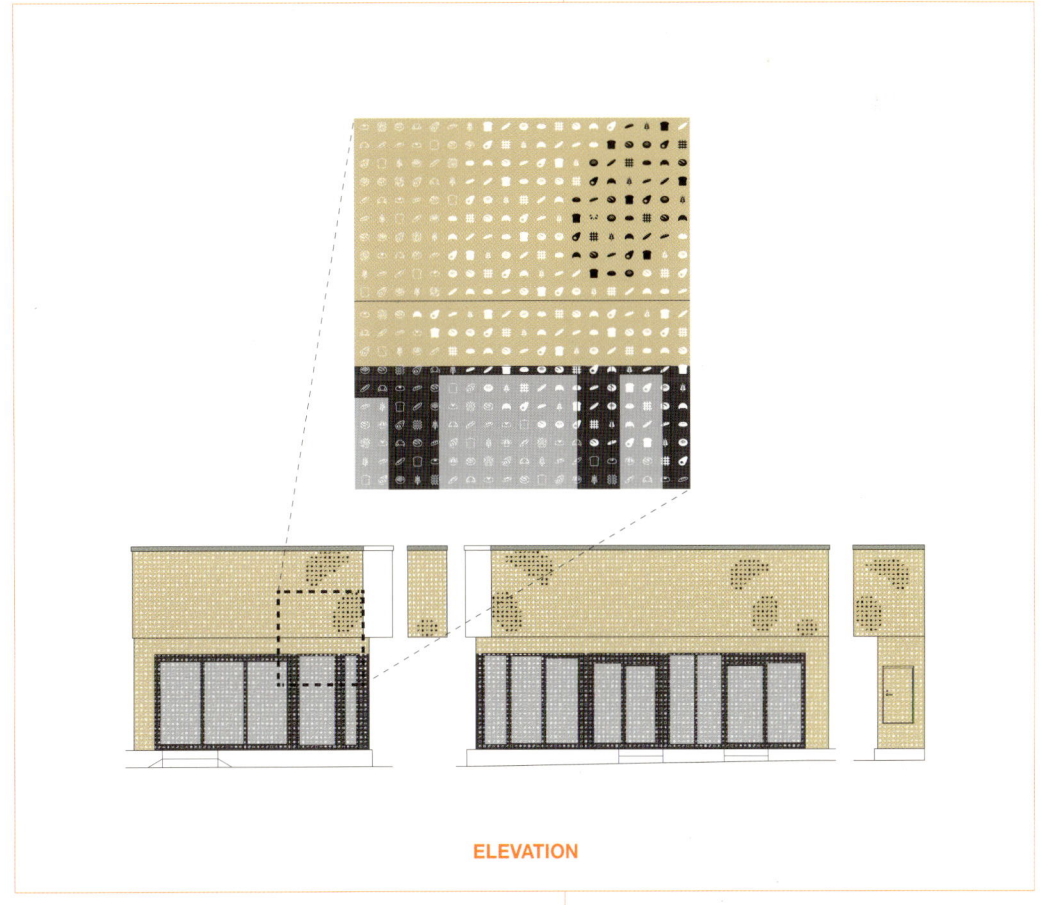

ELEVATION

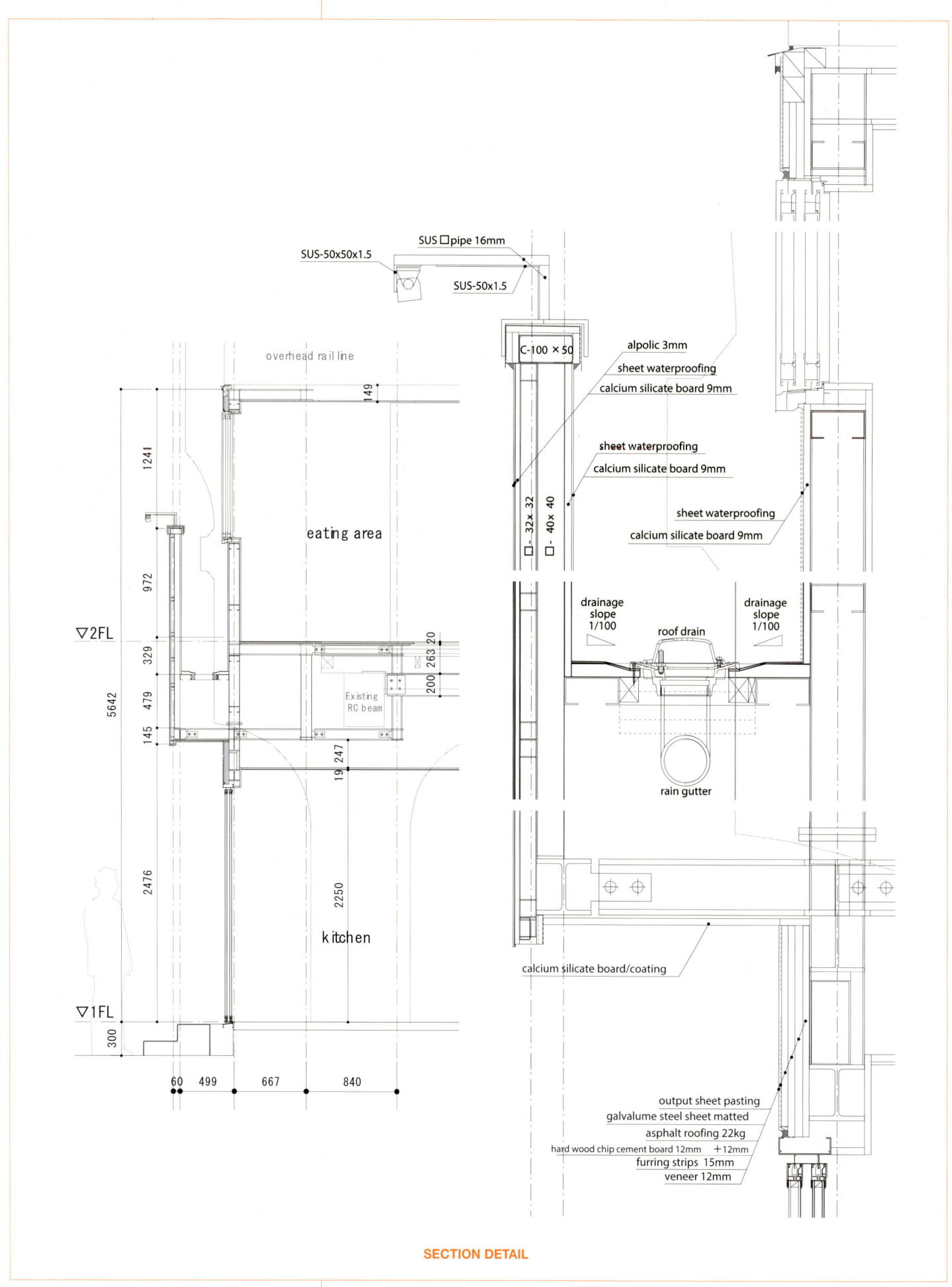

SECTION DETAIL

Neighborhood Facility

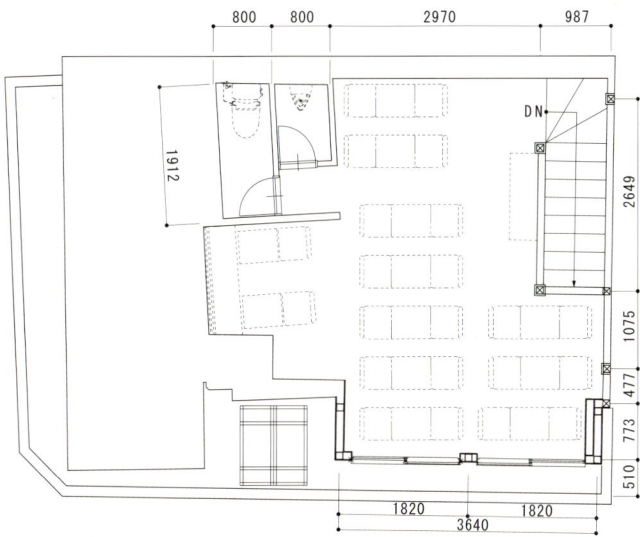

2ND FLOOR PLAN

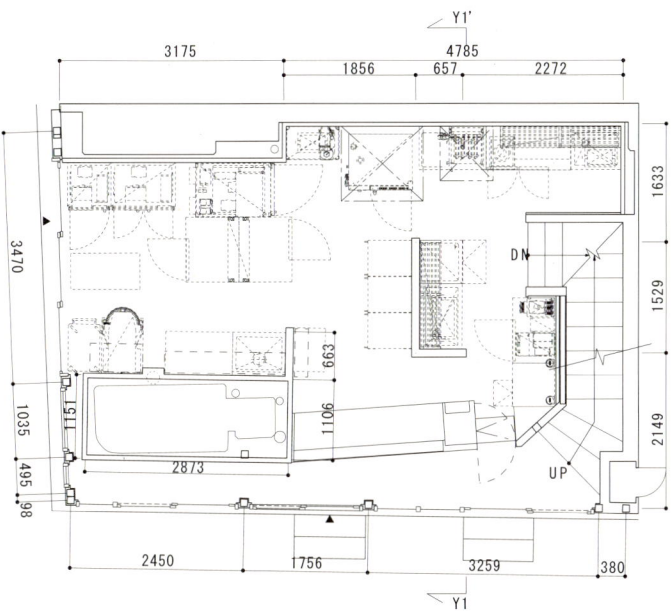

1ST FLOOR PLAN

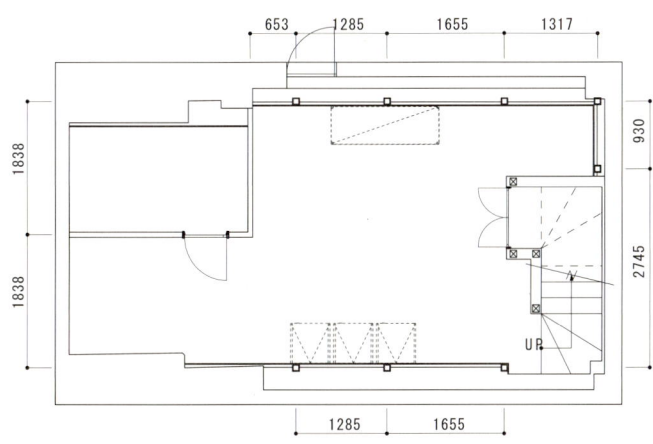

B1 FLOOR PLAN

Keisuke Fukui, Keisuke Morikawa / KAMAKURA STUDIO

Architectural office KAMAKURA STUDIO based in Nagareyama, Chiba, Japan. In recent years, our office has received the 45th Tokyo Architecture Award Newcomer Award and the Good Design Award 2022 for two residences.

The first floor of one of them is KAMAKURA STUDIO. They also have a café in the office, as they want it to be an open architectural office where anyone can feel free to drop by. They operate the café by making own home-roasted coffee and homemade cakes. It is also a place for local residents to hold events.

In addition to residential projects, they are actively involved in activities that benefit the city, such as designing seasonal event spaces sponsored by Nagareyama City and rebuilding and renovating the city's meeting halls.

The office name, "kamakura" means a Snow Hut. Everyone can participate in its creation, and anyone can enter. Communication is naturally created between the creators and passersby. It changes the impression of the surrounding area and create a friendly and enjoyable landscape. Our goal is to create an architectural office that is like a snow hut.

>> www.kamakurastudio.com

근린생활시설 7

51m² − 100m²

NEIGHBORHOOD FACILITY 7

**TETUSIN DESIGN RE-USE OFFICE /
재구성을 통해 설계한 새로운 사무실**
YHA ARCHITECTS

MIIKE / 미케
ihrmk

ACB LIVING / ACB 리빙
LIGHTHOUSE

MH OFFICE / MH 오피스
GENDAI SEKKEI CO.,LTD.

**ARCHITECTURE AND INTERIOR DESIGN STUDIO OFFICE /
건축&인테리어 디자인 스튜디오 사무실**
Manoj Patel Design Studio

NINE CUBE / 나인큐브
(주)서원건축사사무소
SEO WON Architects Cooperation

JAEDONG PROJECT / HALF & HALF BUILDING
재동 프로젝트 / 반반건축
바운더리스 아키텍츠 BOUNDARIES ARCHITECTS

Location
Fukuoka, Japan
Use
Office, house
Site area
310.71m²
Built area
58.82m²
Total floor area
117.27m²
Floors
2F

Exterior finish
Roof_ tent membrane (fiberglass membrane), Wall_ existing cedar boards, ceramic boards + water-based acrylic resin paint, Exterior_ wood deck, sandy soil, existing brick, existing stone
Interior finish
Floor_ slab, solid flooring, long vinyl chloride sheet, Wall_ rigid plasterboard, calcium silicate board, Ceiling_ tent membrane (fiberglass membrane), rigid plasterboard

Project architect
yHa architects / yujin HIRASE + yuko HIRASE
Photographer
Yousuke Harigane, Seinosuke Kaneda

TETUSIN DESIGN RE-USE OFFICE

재구성을 통해 설계한 새로운 사무실

YHA ARCHITECTS

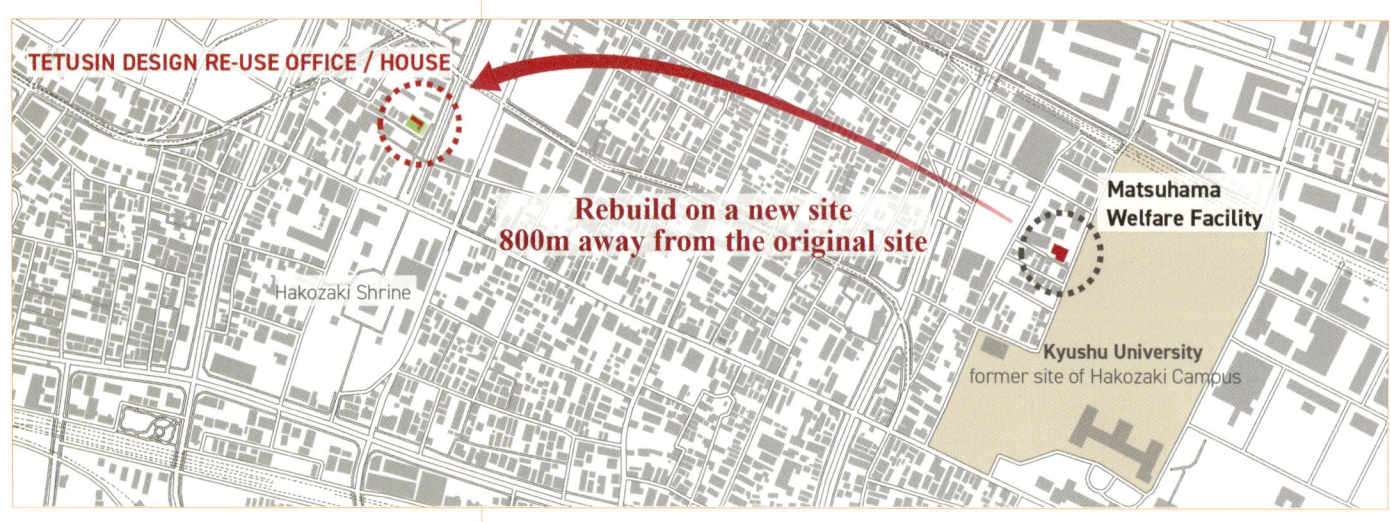

SITE PLAN

Neighborhood Facility

New office / house design by capturing and reconstructing parts and materials

This is a designer's residence with office facing the approach to Hakozaki Shrine in Fukuoka, Japan. Kyushu University Hakozaki campus was about to be demolished, and the university decided to take over the components of the "Kyushu University Matsuhama Welfare Facility" (completed in 1928), a Western-style building. The client of designer, who has long been involved in the preservation and utilization of historical buildings on the former campus site, was concerned that demolition would cause people to lose their memories of Kyushu University here in Hakozaki, which was the impetus for the project. The project is not a strict repair based on traces, as in the case of so-called "cultural property conservation," but rather a "spolia" type of inheritance of memory, in which historical values are connected to new architecture through selective diversion of historical elements

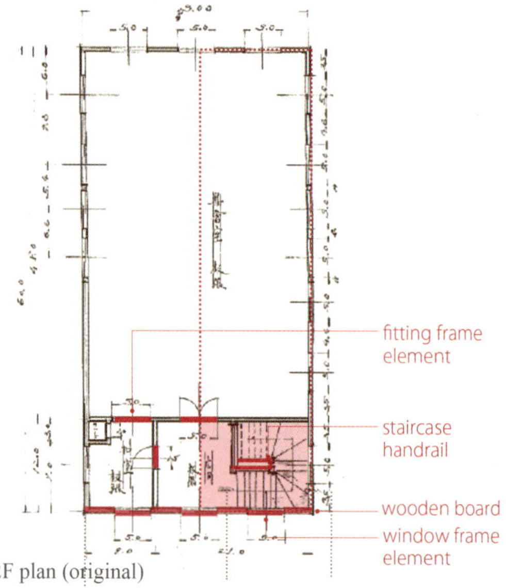

2F plan (original)

- fitting frame element
- staircase handrail
- wooden board
- window frame element

Kyushu University Matsuhama Welfare Facility

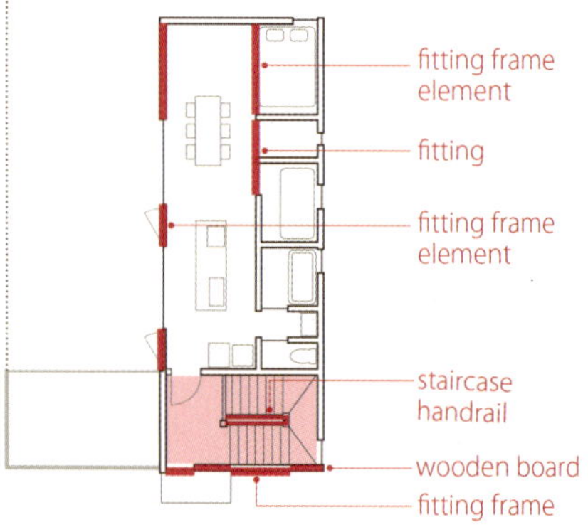

2F plan

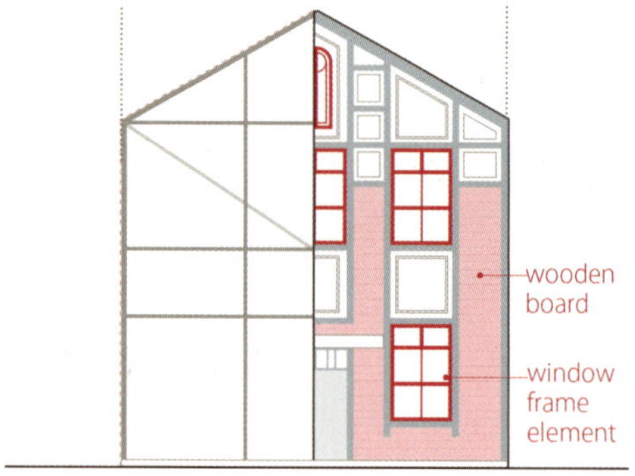

- fitting frame element
- fitting
- fitting frame element
- staircase handrail
- wooden board
- fitting frame

- wooden board
- window frame element

TETUSIN DESIGN RE-USE OFFICE / HOUSE

1 FITTING FRAME ELEMENT
2 FITTING
3 WINDOW FRAME ELEMENT
4 WOODEN BOARD
5 VENTILATION DUCT ORNAMENT

COMPONENTS & MATERIALS DIAGRAM

Neighborhood Facility

BEFORE DEMOLITION

CONSTRUCTION

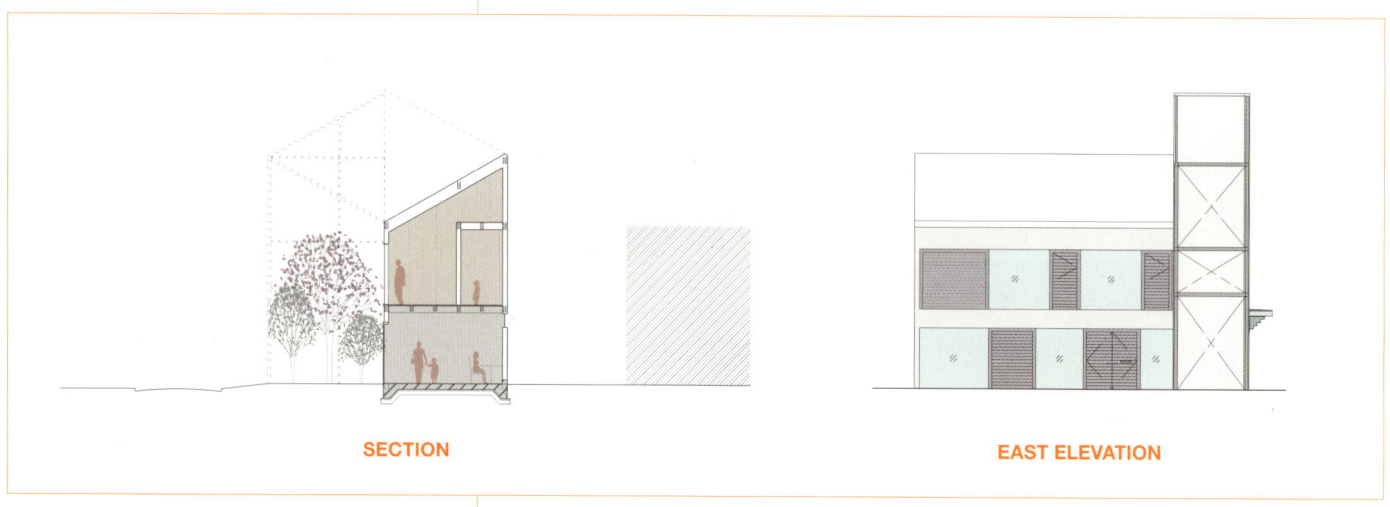

SECTION

EAST ELEVATION

Neighborhood Facility

89

DIAGRAM

Neighborhood Facility

(spolia is the act of reusing elements such as sculptures and columns in Western architecture by diverting them to another building). The juxtaposition and layering of time differences, such as the placement of old fitting frames next to new building materials, results in the acquisition of multilayered meanings. This is an attempt to pass on memory through objects.

In reconstructing the components on a new site approximately 800m away from the original site, we considered it necessary to inherit 3 important memories while maintaining harmony with the surrounding environment and adapting to site conditions such as a quasi-fire prevention zone. The "urban memory" of the symbolic façade, which has become part of the cityscape, the "spatial memory" of the stairwell, which conveys a sense of grandeur with its large volume, and the "material memory" of the building, such as its distinctive façade color and strong fitting frames. The new façade is a contrasting composition of symmetry between mass and void, with a steel frame framework that reinterprets the volume of the historical building as "reality" and "emptiness" as void. The steel frame, which recalls the existing form of the building, ensures a 3-dimensional landscape through greening and the ability to expand in response to changes in the environment of the residents.

The tent-membrane roof fills the interior with a soft quality of light, while outside, it becomes a new symbol of the city, gently diffusing light like an andon lantern. During the festival season, the white tented stalls lined up along the approach to Hakozaki Shrine in front of the building create a sense of unity with the approximately 500 stalls lined up along the approach to Hakozaki Shrine.

1 OFFICE
2 ENTRANCE HALL
3 DINING/KITCHEN
4 BEDROOM
5 WIC
6 CHILDREN ROOM
7 STAIR HALL

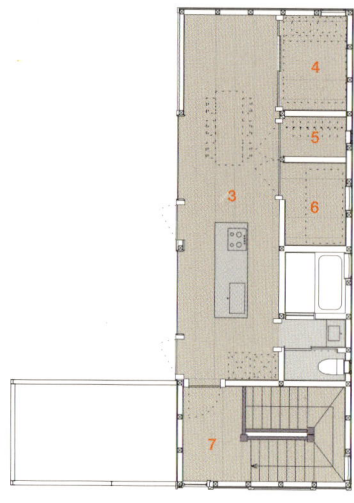

2ND FLOOR PLAN

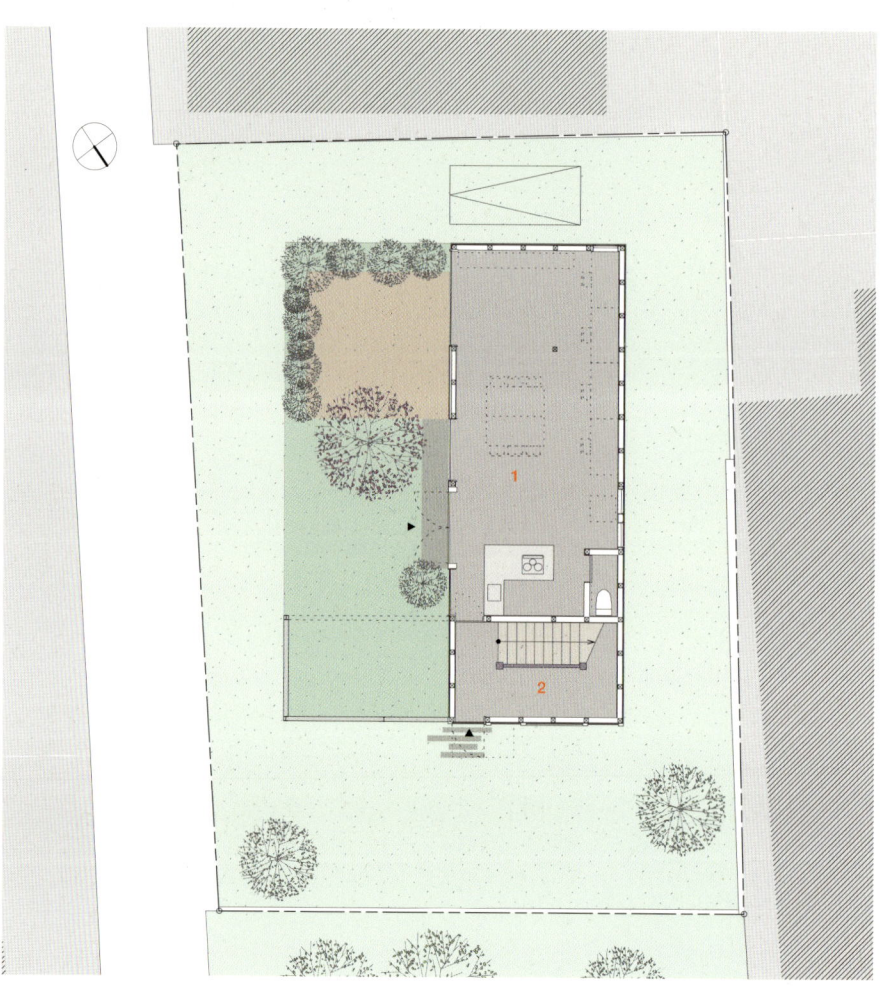

1ST FLOOR PLAN

Neighborhood Facility

부품·자재 수집 및 재구성을 통해 설계한 새로운 사무실 겸 주택 이곳은 일본 후쿠오카 하코자키구 접근로를 마주보고 지은 디자이너의 주거용 사무실이다. 규슈대학 하코자키 캠퍼스 철거가 눈앞으로 다가오자 규슈대학은 서양식 건축물인 "규슈대학 마츠하마 복지시설 (1928년 준공)"의 일부를 인수하기로 결정했다. 이번 고객은 디자이너로서 이전 캠퍼스 부지에 세운 유서 깊은 건물의 보존 및 활용에 오랫동안 관여해 왔으며, 철거가 진행되면 사람들이 본 프로젝트의 원동력이 된 이곳 하코자키의 규슈대학에 대한 추억을 잃을 것을 우려했다. 본 프로젝트는 이른바 "문화재 보존"처럼 흔적에 기반한 철저한

보수가 아니라, 역사적 요소를 선별적으로 전환하여 새로운 건축물에 역사적 가치를 연결하는 "스폴리아" 방식의 기억에 대한 계승이다(스폴리아는 서양 건축에서 조각이나 기둥과 같은 요소를 다른 건물에 적용하여 재사용하는 방식). 새로운 건축 자재 옆에 낡은 고정 프레임을 배치하는 등 시간차의 병치와 중첩은 결과적으로 다중적인 의미를 부여한다. 이는 객체를 통해 기억을 전달하려는 시도이다.

기존 부지에서 약 800미터 떨어진 신설 부지에 진행하는 재건축은 주변 환경과의 조화, 준화재방지구역 등 부지 여건에 적응하는 동시에 세 가지 중요한 기억을 전달해야 한다고 판단했다. 도시 경관의 일부로서 상징적 파사드의 "도시적 기억", 커다란 부피로 웅장한 느낌을 전달하는 계단통의 "공간적 기억", 마지막으로 독특한 파사드 색상과 강력한 고정 프레임 같은 빌딩의 "물질적 기억"이 그 세 가지 기억이다. 새로운 파사드는 역사적 건물의 부피를 "현실"로, 빈 공간을 "공허함"으로 재해석한 스틸 프레임 골조로서 물질과 빈공간의 대조되는 대칭 구조를 이룬다. 빌딩의 기존 형태를 연상시키는 스틸 프레임으로 녹지 조성을 통해 입체적인 경관이 확보되고 거주자의 환경 변화에 대응한 가능성이 확장된다.

천막형 지붕은 내부를 부드러운 빛으로 채우고 외부는 안돈(andon) 랜턴처럼 은은한 빛을 발하여 도시의 새로운 상징이 된다. 축제 기간에는 건물 앞 하코자키 신사로 가는 길을 따라 쭉 늘어선 하얀 천막 노점들은 신사 참배길에 늘어선 약 500개의 노점들과 일체감을 형성한다.

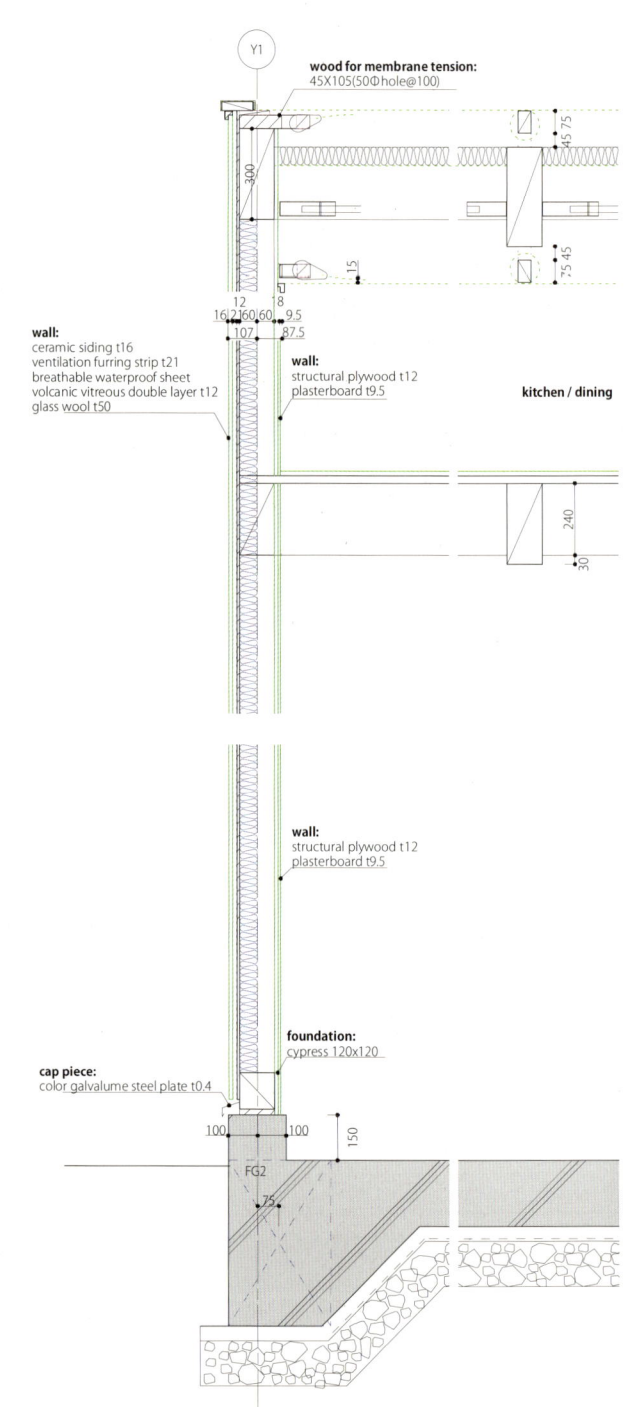

Neighborhood Facility

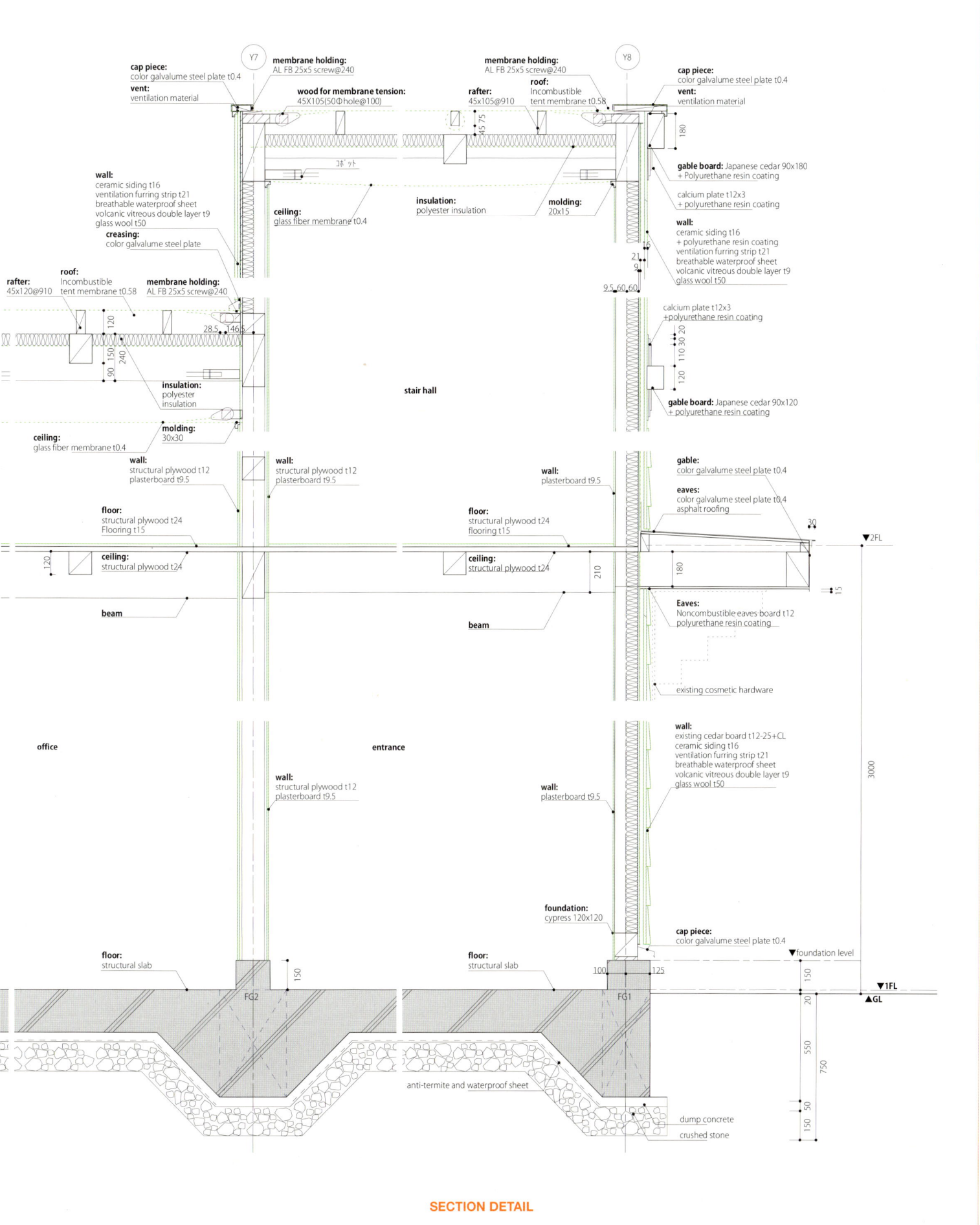

SECTION DETAIL

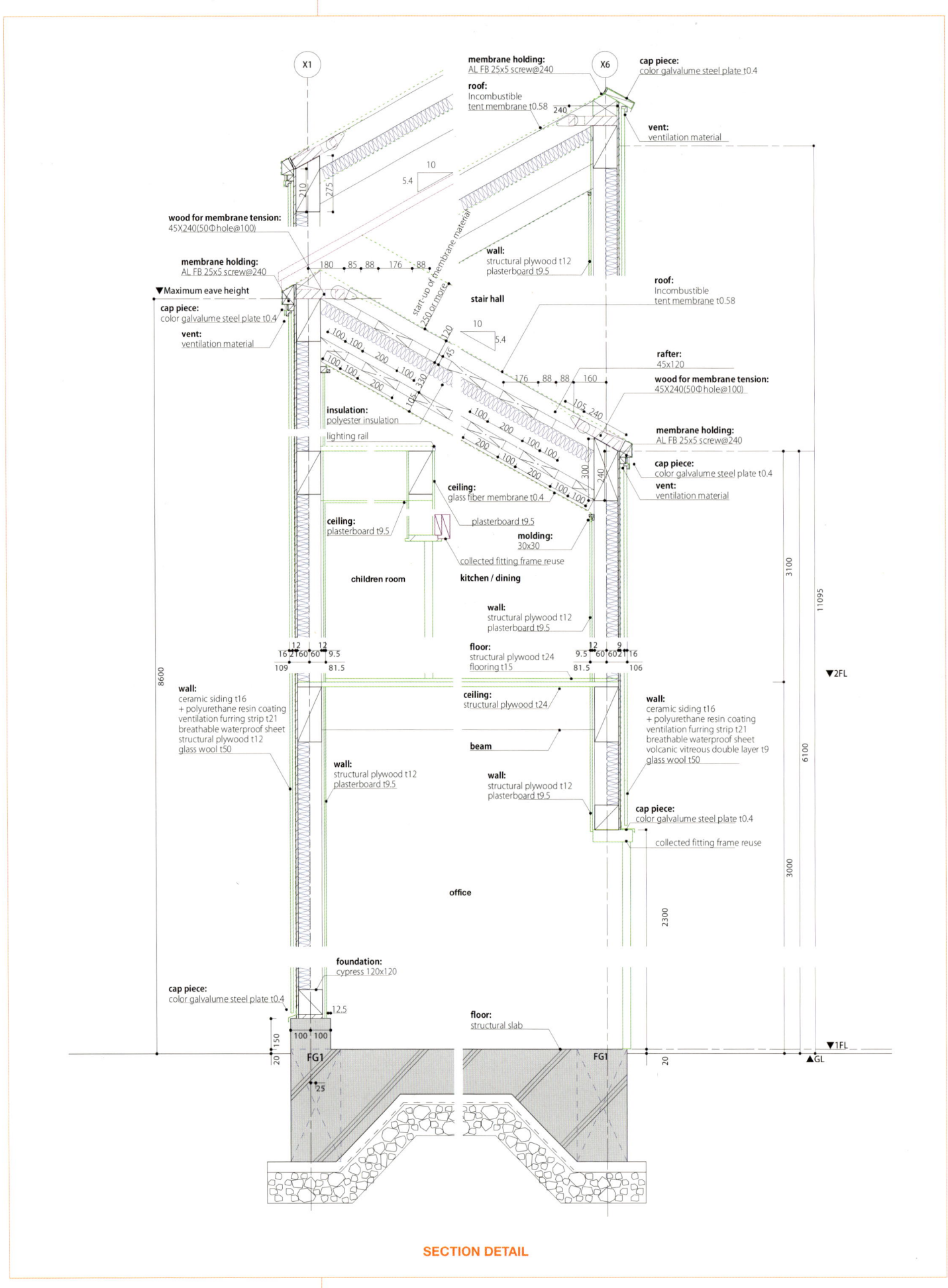

SECTION DETAIL

Neighborhood Facility

yujin HIRASE + yuko HIRASE / yHa architects

yujin HIRASE

yujin HIRASE was born in Tokyo in 1976, he received Dr. Arch., M. Arch. and B. Arch. from Waseda University. After his graduation he had been an assistant at Waseda University and collaborations with Prof. Nobuaki Furuya and NASCA. He established yHa architects since 2007 and worked in Switzerland as a Japanese Government (Agency for Cultural Affairs) Overseas Study Program for Artists during 2007-08. Since 2008 he is an Associate Professor at Saga University.

yuko HIRASE

yuko HIRASE was born in Tokyo in 1975, she received M. Arch. and B. Arch. from Waseda University. After her graduation she worked at Fujie Kazuko Atelier and collaborations with HHF architects in Switzerland. Since 2008 she has co-founded yHa architects.

They have also received numerous awards including Selected Architectural Designs 2017 / 2021-22 (Architectural Institute of Japan), the Award for Excellent Architecture 2015 / 2019 (The Japan Institute of Architects), JIA AWARD Grand Prix 2016 (Japan Interior Designers' Association), Good Design Award 2016 / 2020 / 2022, SD Review 2014 / 2019, 4th ADAN (Architectural Design Association of Nippon) Prize 2021.

\>> yha.jp

Location
Nagano, Japan
Use
Hair salon
Site area
247m²
Built area
64.83m²
Total floor area
64.83m²

Floors
1F
Project architect
IHARA Masaki + IHARA Kayo, ONO Kojiro
Photographer
CHIBA Kenya

miike

미케

ihrmk

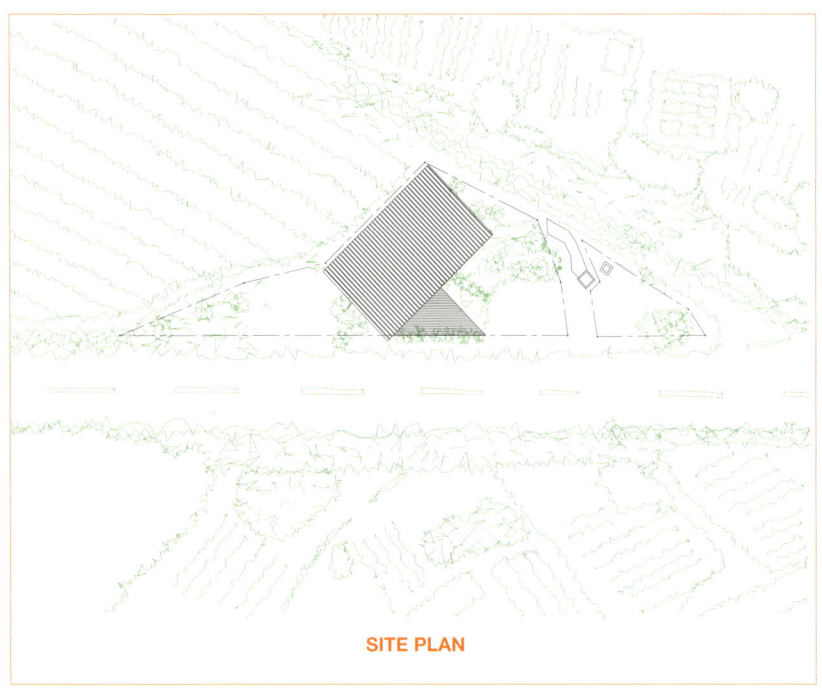

SITE PLAN

Neighborhood Facility

BEFORE

New but familiar: merging into the rural scenery

"miike" is a conversion project for a hairstylist couple who returned to their hometown at the foot of the Asama mountain after spending more than 15 years in Tokyo. The couple eyed an existing prefab structure used to store farming equipment, which now houses their new hair salon and a photo studio.

The site is located on an unpaved path connecting a small Torii gate and an abandoned shrine that overlook the town of Tomi and the Yatsugatake mountains. Here, the vast fields and scattered prefabs, much like miike, are forming a pastoral scene.

One of the challenges of the project was how to insert a new program into an established local community without disrupting the existing scenery. These prefab structures

Neighborhood Facility

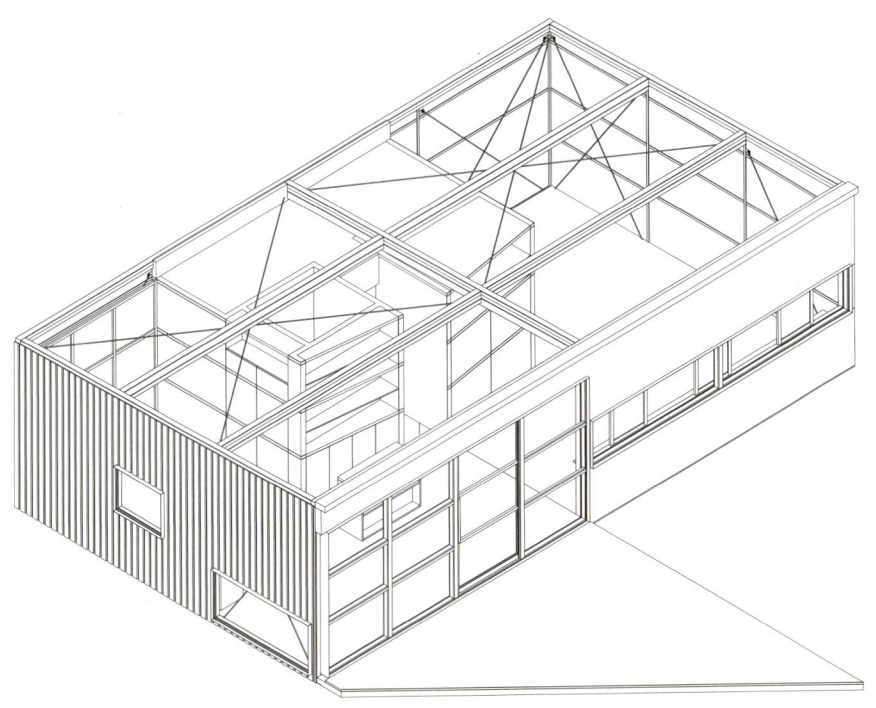

AXONOMETRIC

have been familiar elements to the locals and are an essential architectural typology that forms the area's character. We determined that converting this 25-year-old prefab was an appropriate choice for this project.

Re-utilizing an ordinary prefab structure

The existing prefab structure stood on a 10 m wide and 6 m long concrete foundation. The entire south-facing side could be opened up to the field in front of it by rolling up the shutter. The other three sides were covered with rectangular corrugated galvanized steel sheets on steel furring strips. It was a simple structure built solely to function as farming equipment storage.

In this project, the furring strips, formerly supporting the exterior walls, are used as a design code to accentuate windows, walls and furniture. The arrangement of openings was determined based on the spacing between the furring strips and columns. The walls and furniture have white lines that match the furring strips' height or width to add another dimension to their presence.

Taking in small cues from outside

The salon's interior and exterior are designed to frame various features that came with the structure. Movable double-sided mirrors are set to reflect the surrounding mountains. The floorboards of the salon and the deck are laid parallel to the adjacent road, while the floor-level window captures a section of the path to the shrine. These mutual references between the elements inside and outside as well as old and new were incorporated to create this new but familiar space.

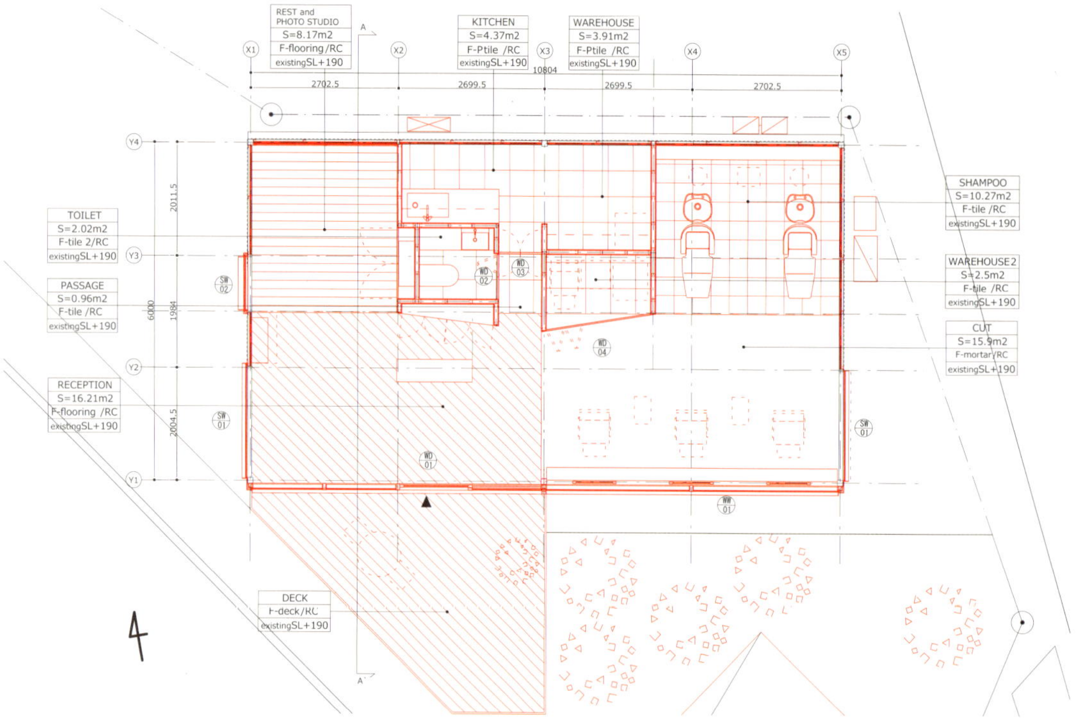

PLAN

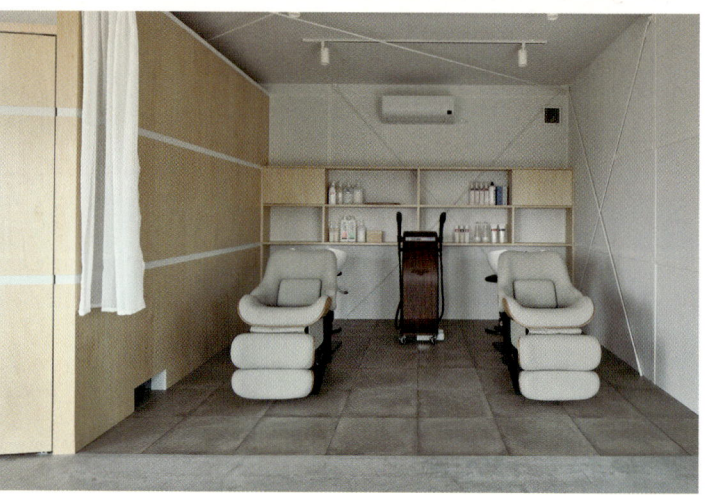

Neighborhood Facility

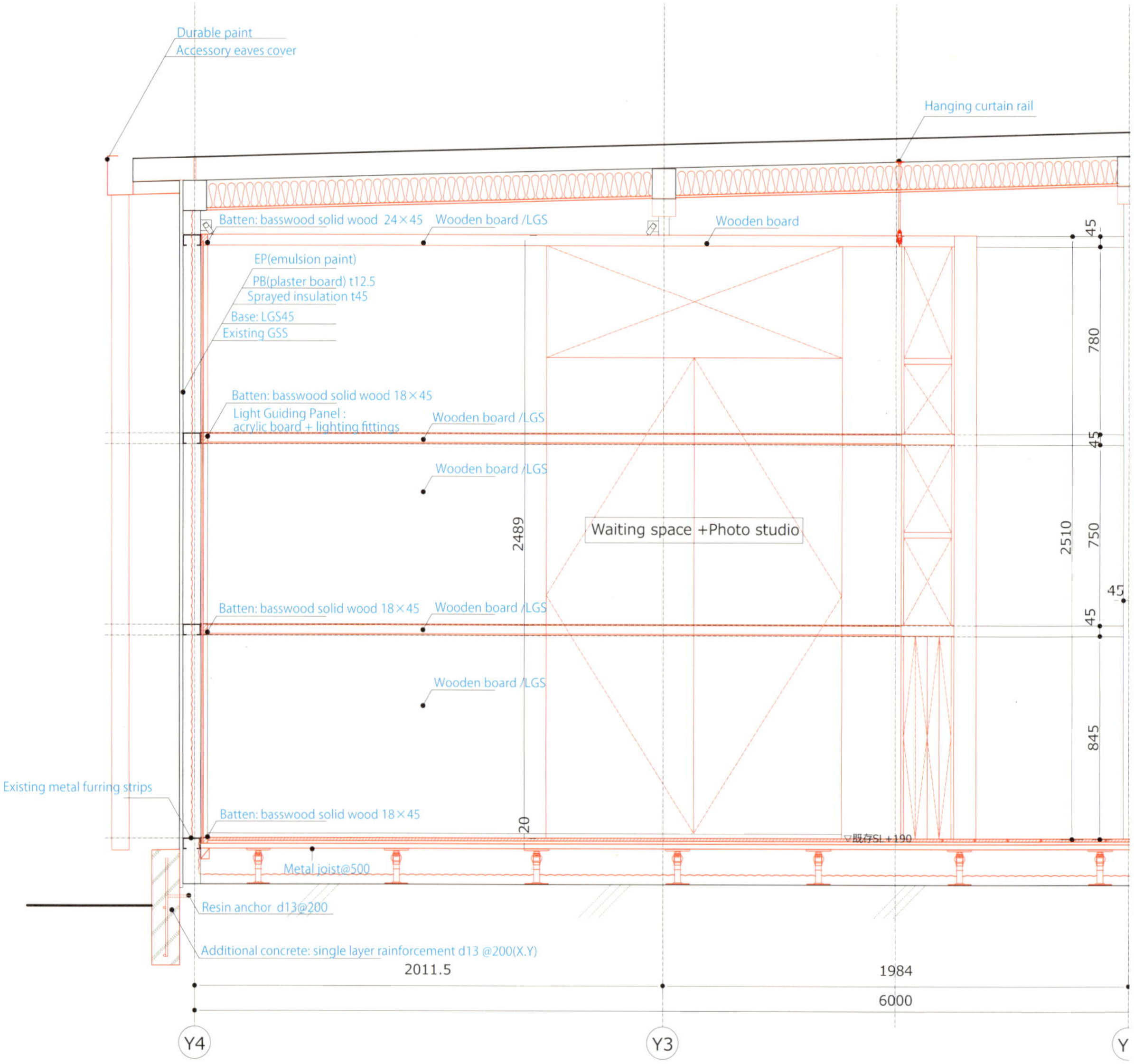

SECTION DETAIL

- Durable paint
- Accessory eaves cover
- Hanging curtain rail
- Batten: basswood solid wood 24×45
- Wooden board /LGS
- Wooden board
- EP (emulsion paint)
- PB (plaster board) t12.5
- Sprayed insulation t45
- Base: LGS45
- Existing GSS
- Batten: basswood solid wood 18×45
- Light Guiding Panel : acrylic board + lighting fittings
- Wooden board /LGS
- Wooden board /LGS
- Waiting space + Photo studio
- Batten: basswood solid wood 18×45
- Wooden board /LGS
- Wooden board /LGS
- Existing metal furring strips
- Batten: basswood solid wood 18×45
- Metal joist@500
- Resin anchor d13@200
- Additional concrete: single layer rainforcement d13 @200(X.Y)

Neighborhood Facility

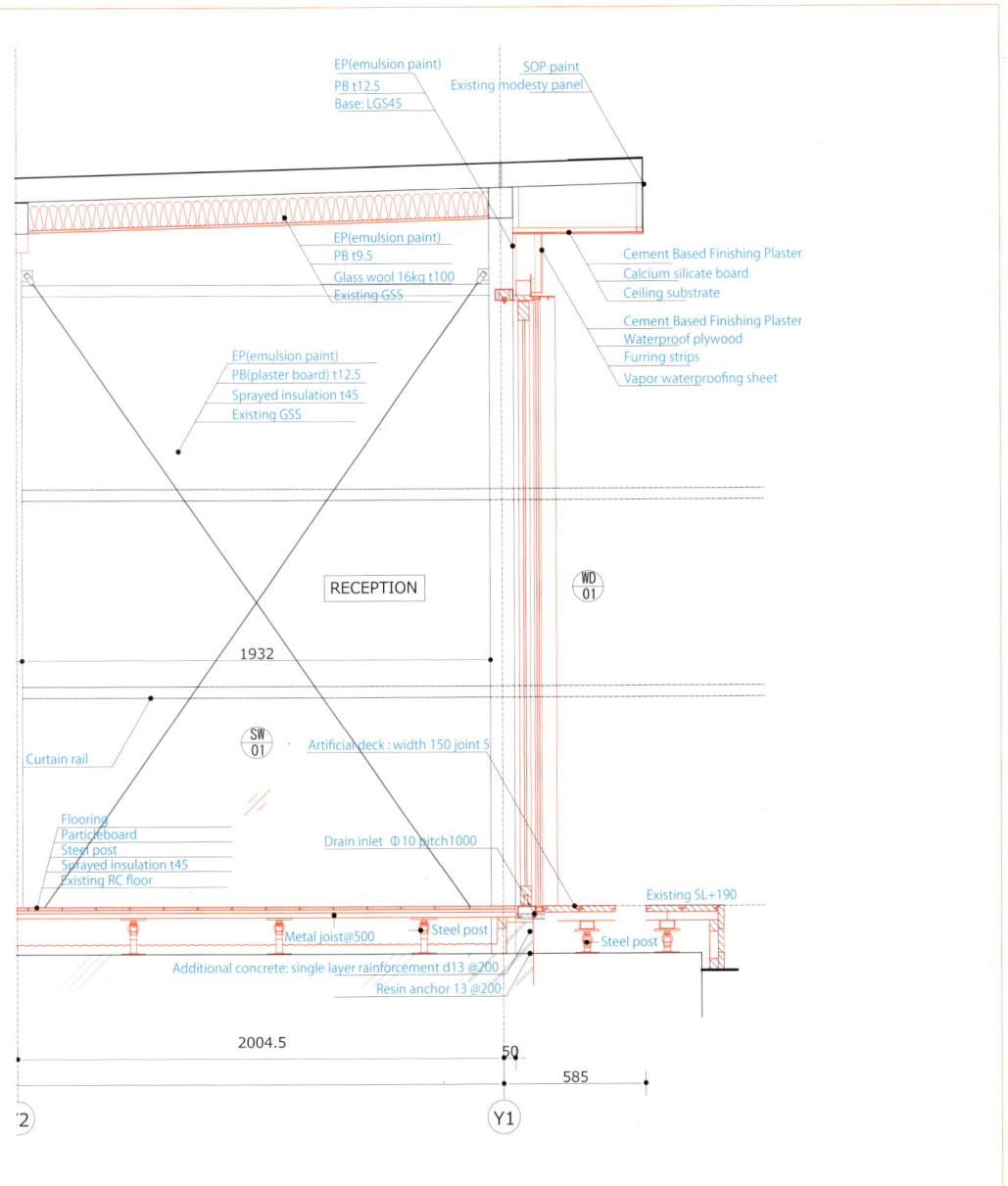

새롭지만 익숙하게: 시골 풍경과의 조화

"miike"는 15년 이상의 도쿄 생활을 청산하고 아사마 산기슭 고향으로 돌아온 미용사 부부를 위한 전환 프로젝트이다. 이 부부는 농기구를 보관하는 기존의 조립식 건물에 주목했으며 이제 이 건물은 새로운 미용실과 사진 스튜디오로 재탄생했다.

건물 부지는 다케토미 마을과 야쓰가타케산이 내려다 보이는 작은 토리이 문과 버려진 신사를 연결하는 비포장 도로에 있다. 이곳의 광활한 들판과 흩어진 조립식 건물은 miike와 마찬가지로 목가적인 장면을 연출한다.

본 프로젝트에서 어려웠던 부분은 기존 경관을 해치지 않고 마을에 새로운 프로그램을 도입하는 방식이었다. 이러한 조립식 구조물은 주민들에게 친숙한 것이었으며 이 지역의 특성 중 하나인 필수적인 건축 유형이다. 이 25년 된 조립식 건물을 개조하는 작업이 본 프로젝트를 위한 적절한 선택이라고 판단했다.

일반 조립식 구조물의 재사용

기존 조립식 구조물은 폭 10m, 길이 6m의 콘크리트 기초 위에 세워졌다. 남향면 전체는 셔터를 말아올려 앞 들판 쪽으로 열 수 있다. 나머지 3면은 강철 띠장 위에 직사각형의 주름진 아연 도금 강판으로 덮여 있다. 농기구 보관소 용도로만 사용하기 위해 지은 단순한 구조였다.

본 프로젝트에서 이전에 외벽을 지탱하던 띠장은 창, 벽, 가구를 부각시키는 디자인 요소로 사용했다. 출입구 배치는 띠장과 기둥 사이의 간격을 기준으로 결정했다. 벽과 가구에는 띠장의 높이 또는 너비와 일치하는 흰색 선이 있어 또 다른 차원이 느껴진다.

외부에서 들어오는 작은 신호

미용실의 내외부는 다양한 기능을 구성하도록 구조적으로 설계했다. 이동식 양면 거울은 주변 산을 비추도록 설치했다. 미용실의 마루판과 데크는 인접한 도로와 평행을 이루고 바닥 높이의 창을 통해 신사로 가는 길이 일부 바라보이도록 했다. 오래된 것과 새로운 것, 내부 요소와 외부 요소를 상호 참조하여 새롭지만 익숙한 공간을 창조해 냈다.

BEFORE

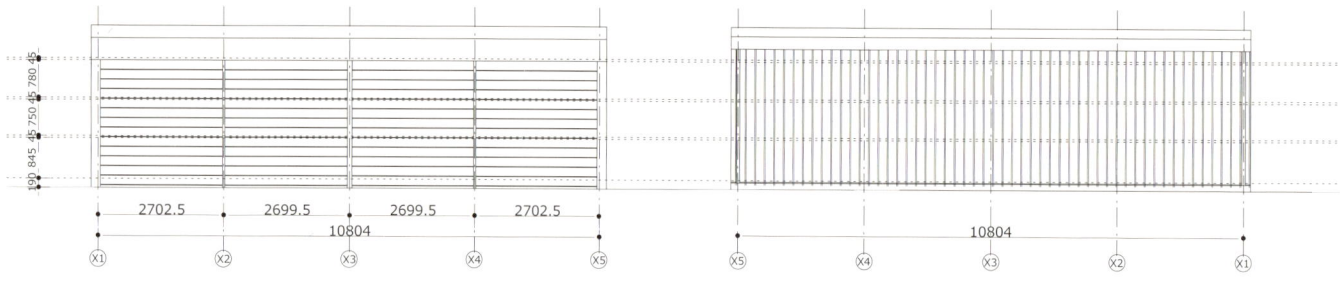

AFTER

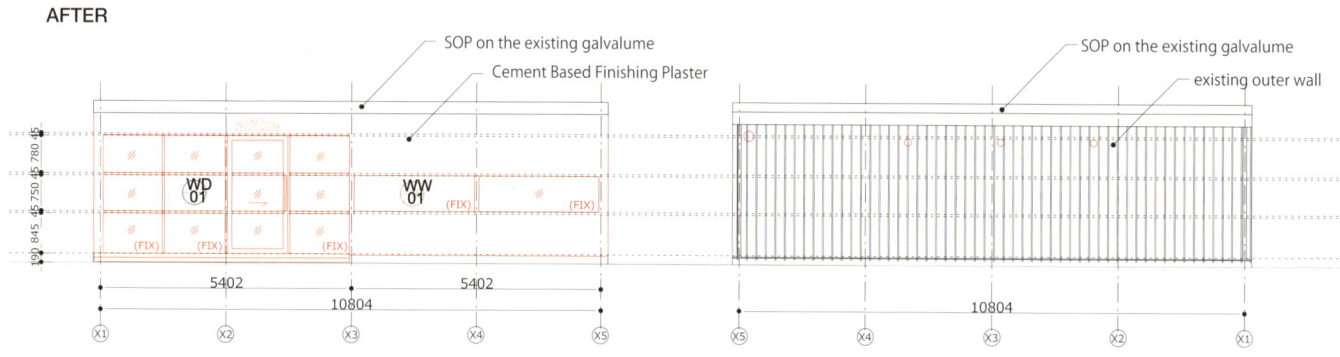

ELEATION DETAIL

Neighborhood Facility

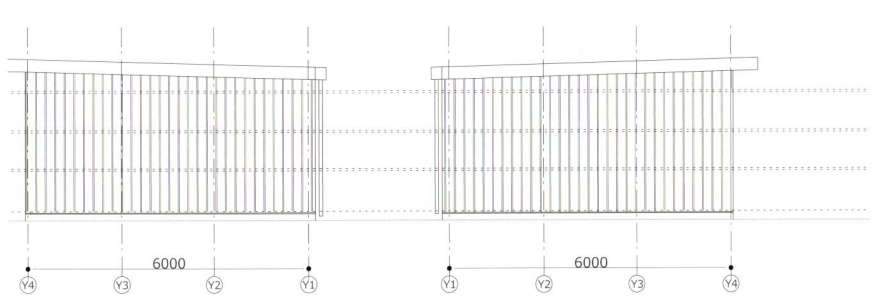

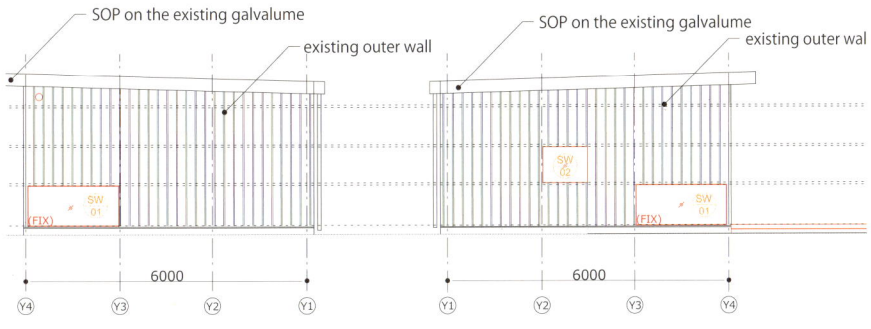

IHARA Masaki + IHARA Kayo/ ihrmk

ihrmk is a architectural design firm by Masaki Ihara and Kayo Ihara, based in Minato-ku Tokyo and Toyota City, Aichi Prefecture Japan. we design and supervise from new constructions, renovations to funiture.

In our design activities, we are conscious of "architecture in situation/design in the situation". we are finding a new way of architecture by designing the situation itself from dynamic situations, responding to changes in physical, functional and social situations, and feedback between the situation and the design.

>> ihrmk.co.jp

Location
Iki Island, Nagasaki, Japan
Use
Coworking space, café
Site area
112.45m²
Built area
72.64m²
Total floor area
108.46m²
Floors
1F, 2F

Exterior finish
Reinforced fiber cement siding panels
Interior finish
Mortar, poured colored concrete, structural plywood, wallpaper
Project architect
Tatsuhiro Shinozaki, Kentaro Hayashi
Photographer
YASHIRO PHOTO OFFICE

ACB Living

ACB 리빙

LIGHTHOUSE

The project is located on Iki Island, a small remote island in Nagasaki Prefecture, Japan. The program of ACB Living consists of a co-working space and its front desk, off-site training spaces for companies, remote work rooms and a café.

By scattering the program across the town, the users are invited to stroll around the area. This way, accidental encounters with the residents of the town can naturally occur and spontaneous conversations ensue. Instead of gathering the whole entity into one single building, the program is distributed across town with new construction and refurbished small scale interventions. Such a scope is a better fit with the town's scale, allowing the neighbors to feel more familiar to these new spaces.

The town has long been suffering from the vacant houses and empty plots problem. By extending the program of ACB Living throughout the town and letting it take different shapes, the company aims to tackle with the town's pressing problem while growing its premises.

The interior spaces of the first and second floor are reduced to the minimum in order to expand the outdoor space at the back of the plot to its fullest extent and consequently, guarantee low-energy consumption. Taking into consideration the pre-existing conditions surrounding the plot and the region's climate, we explored a passive energy solution which provides an optimal use of the site.

Three sides of the plot are tightly encircled by neighboring buildings. Taking advantage of this peculiar characteristic and relying on site-specific air flow, a wall-less double heigh outdoor space is laid out. In order to control the amount of light and wind exposure, agricultural sheets are placed along the perimeter of the outdoor space which the users can draw freely. The billowing

Neighborhood Facility

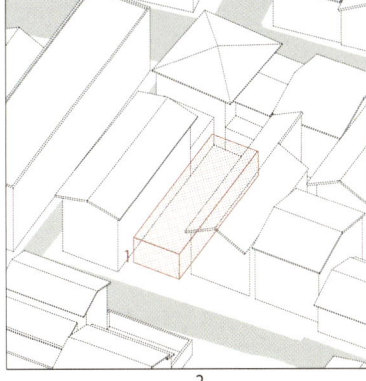
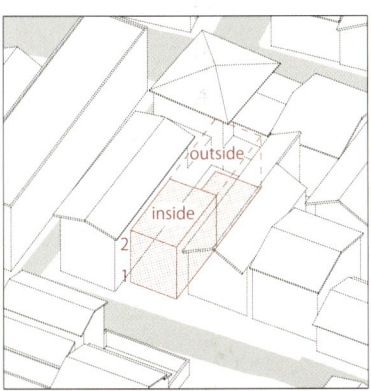
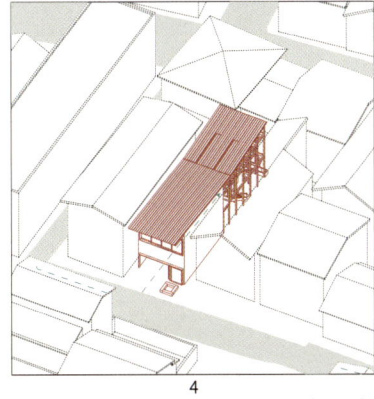

1
Vacant space in a crevice of the town.

2
The limited budget only allowed for a single-storey building with a maximum occupancy of the plot.

3
A two-storey building with a footprint of half of the plot faces the front, while tha back is liberated to implement an outdoor space.

4
In order to bring back a codependent relationship with neighbors, an attractive outdoor space is conceived in the back.

DIAGRAM

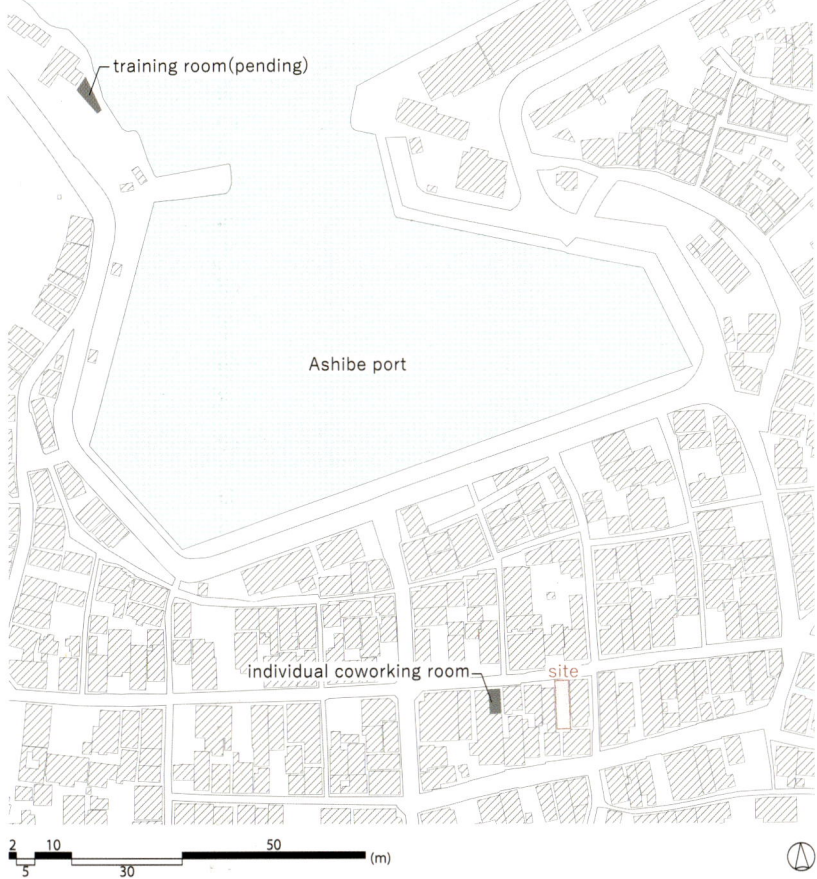

SITE PLAN

agricultural sheets help visualize the presence of the wind and, at the same time, blur the lines of the site's perimeter. Moreover, the previously mentioned neighboring walls become part of the project's interior design.

In the past, a common sight across town was of plants, chairs and other objects whose property was ambiguous. Even the houses' private entryway was loosely connected to the public realm. The project aims to bring back a codependent relationship with its neighbors by creating an appropriate environment where this reliance can flourish.

The whereabouts of this project —located in a crevice on the streets of the town—, unlike in urban areas where only mechanical noise can be heard, offer its users a different kind of noise: the sound of the residents' activities, the birds singing and the insects chirping.

The spatial arrangement that has been tried out for this project, isn't necessarily restricted to its current program and could be easily transformed into a residence or a store. What's more, it's a proposal for the region for a new lifestyle where everyone is loosely connected.

본 프로젝트는 일본 나가사키현의 작은 외딴 섬인 이키섬에 있다. ACB 리빙(ACB Living) 프로그램은 함께 일할 수 있는 공간과 프론트 데스크, 부지 밖의 기업 전용 교육 공간, 원격 작업실과 카페 등으로 구성되어 있다.

본 프로그램을 마을 전체에 분산시켜 사용자들은 그 지역 주변을 산책할 수 있다. 이러한 방법으로 마을 주민들과의 우연한 만남이 자연스럽게 발생하고 이는 자연스러운 대화로 이어진다. 조직 전체를 한 빌딩으로 모으는 대신 본 프로그램은 새로운 건축과 새로 단장한 소규모의 개입을 통해 도시 전체로 분산된다. 이러한 분산된 범위는 마을 규모와 더 잘 맞아 동네 사람들은 이 새로운

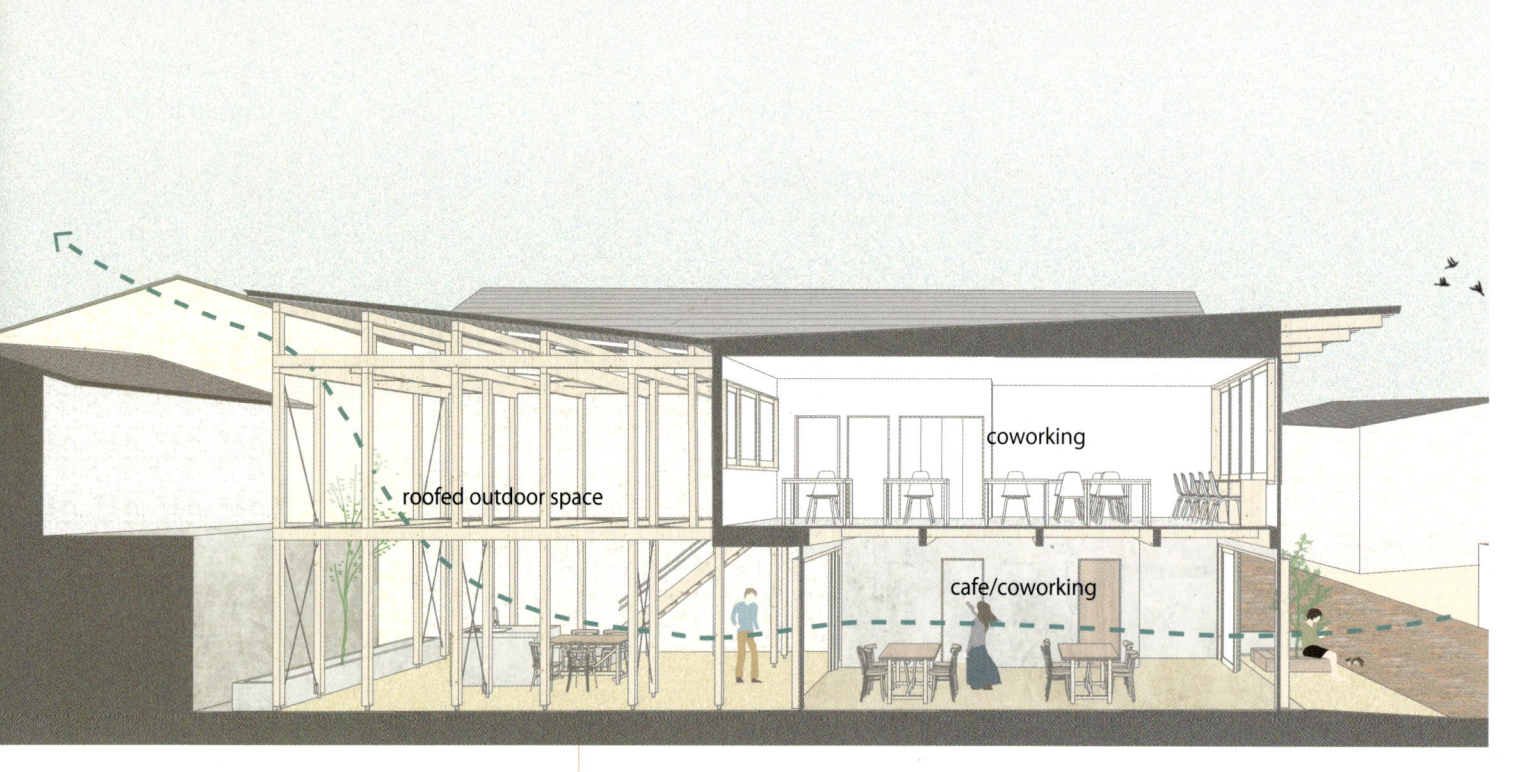

SECTION PERSPECTIVE

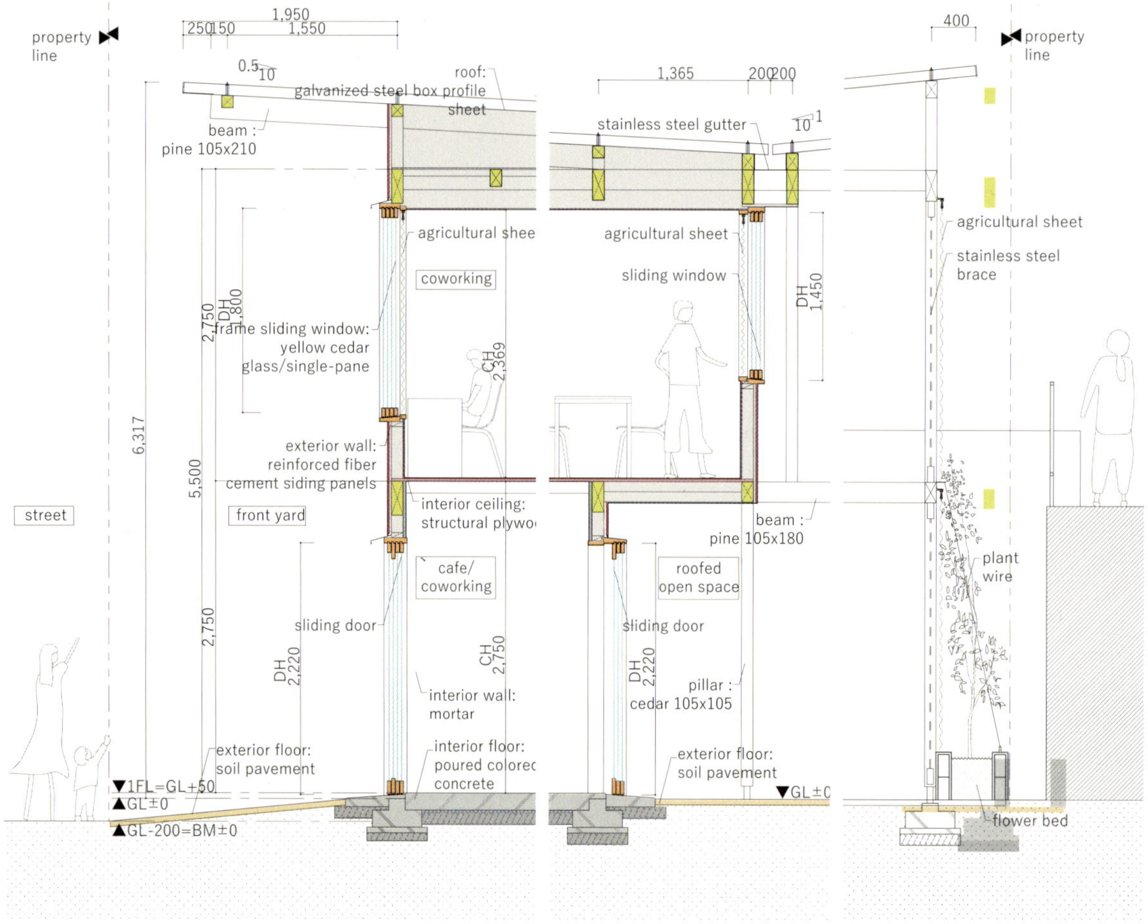

SECTION DETIAL

Neighborhood Facility

1 ROOFED OUTDOOR SPACE
2 CAFE/COWORKING
3 TOILET
4 REMOTE WORK ROOM
5 RECEPTION
6 COWORKING
7 STORAGE

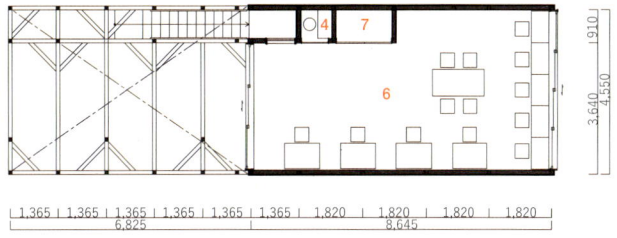

2ND FLOOR PLAN

1ST FLOOR PLAN

공간에 더욱 친숙함을 느끼게 된다.

마을은 오랫동안 빈집과 공터 문제로 골머리를 앓아 왔다. ACB 리빙 프로그램을 마을 전체로 확장하고 다양한 형태를 취하게 함으로써 회사는 부지를 확장하는 동시에 마을의 시급한 문제를 해결하는 것을 목표로 한다.

1층과 2층의 실내 공간을 최소화하고 부지 뒤편의 실외 공간을 최대한 늘려 에너지 소비를 절감한다. 획지(plot)와 지역 기후 등 주변 환경을 고려하여 부지 사용의 최적화를 위한 패시브 에너지 솔루션을 탐구하였다.

획지의 3면은 이웃한 건물들이 빽빽이 둘러싸고 있다. 이러한 독특한 특성을 활용하고 부지별 공기 흐름에 기반하여 벽이 없는 2층 높이의 실외 공간을 만든다. 빛과 바람의 노출량 조절을 위해 농업용 시트는 야외 공간의 둘레를 따라 배치한다(사용자가 자유롭게 정할 수 있음). 바람에 휘날리는 농업용 시트는 바람의 존재를 시각화하는 데 도움이 되는 동시에 현장 주변의 경계를 흩트려 놓는다. 앞서 언급한 이웃해 있는 벽들은 본 프로젝트 내부 디자인의 일부가 된다.

과거 마을에는 식물, 의자, 소유주가 불분명한 건물들이 자주 눈에 띄었다. 심지어 그 집들의 비공개 출입구는 공공 영역과 느슨하게 연결되어 있었다. 본 프로젝트는 서로 의지할 수 있는 적절한 환경을 조성함으로써 이웃과의 상호 의존 관계에 다시 숨을 불어 넣는 것을 목표로 한다. 본 프로젝트는 마을 거리의 틈새에 위치하여 기계 소음만 들리는 도시 지역과는 달리 사용자들에게 주민들의 생활 소리, 새들의 지저귐, 곤충들의 재잘거림 같은 다른 종류의 소리를 제공한다.

본 프로젝트에 시도한 공간 배치는 반드시 현 프로젝트에만 국한되는 것이 아니라 쉽게 주거지나 상점으로 전환될 수 있다. 나아가 이는 사람과의 연결성이 느슨한 새로운 생활방식을 추구하는 지역을 위한 아이디어이다.

Neighborhood Facility

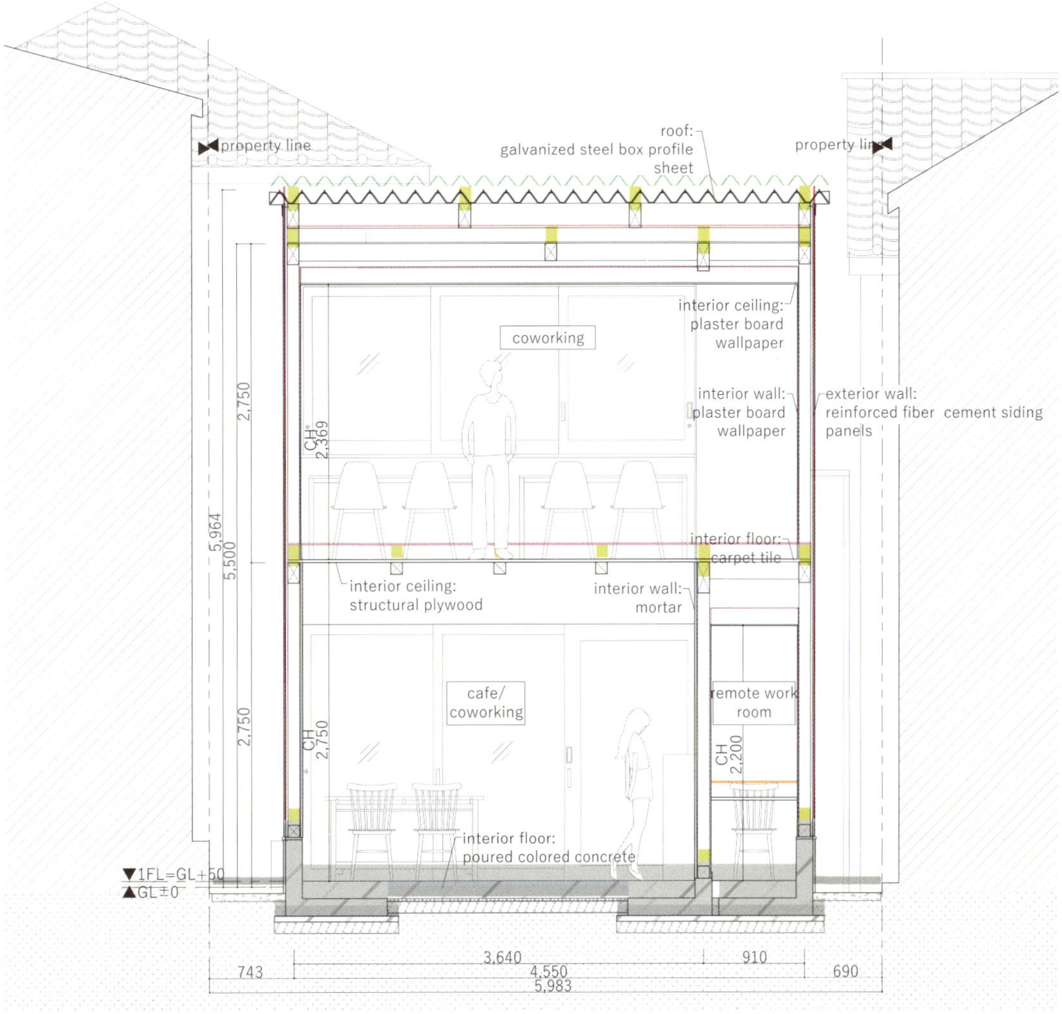

SECTION DETIAL

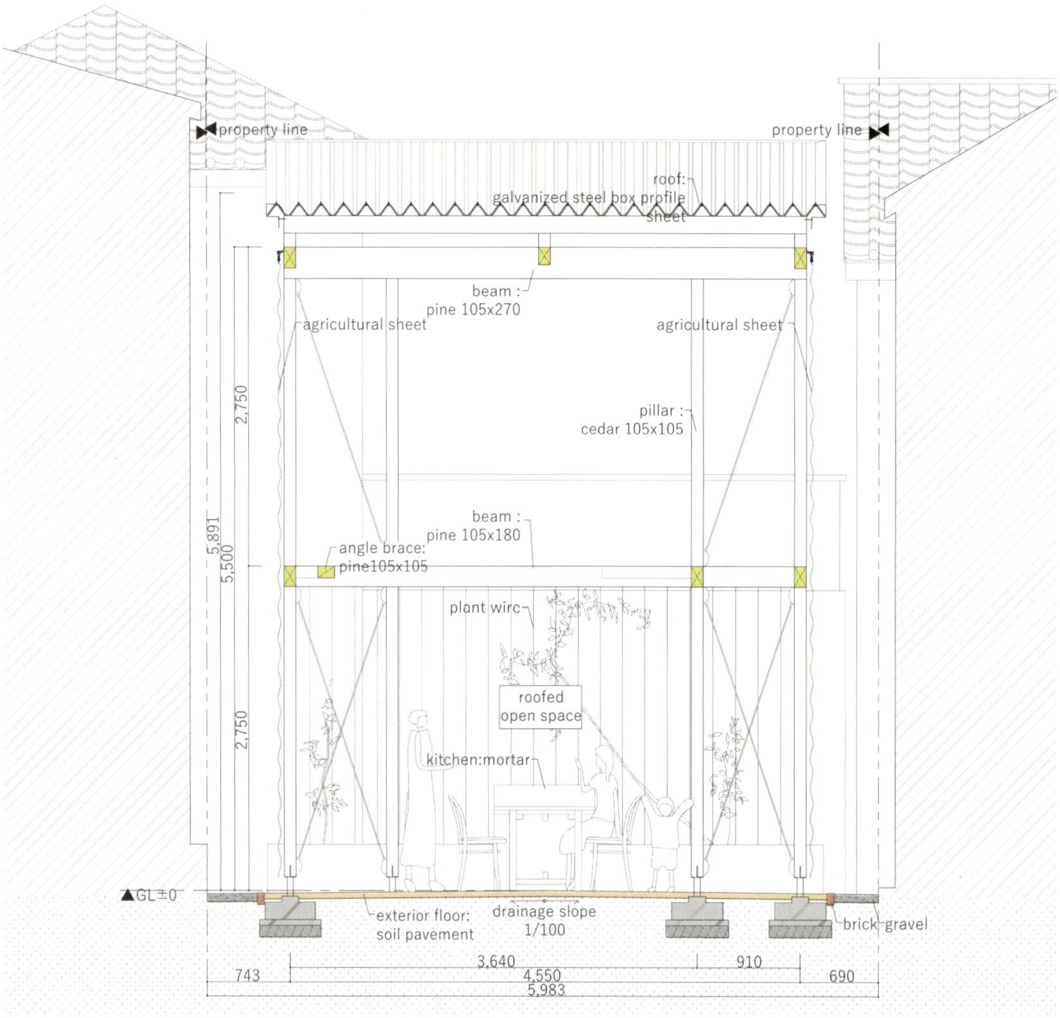

SECTION DETIAL

- roof: galvanized steel box profile sheet
- property line
- beam : pine 105x270
- agricultural sheet
- pillar : cedar 105x105
- beam : pine 105x180
- angle brace: pine105x105
- plant wire
- roofed open space
- kitchen:mortar
- exterior floor: soil pavement
- drainage slope 1/100
- brick gravel
- ▲GL±0

Dimensions: 2,750 / 2,750 / 5,500 / 5,891
743 / 3,640 / 910 / 690
4,550
5,983

Neighborhood Facility

Tatsuhiro Shinozaki / LIGHTHOUSE

We are based on Iki Island, a small remote island in Japan, where we run a design practice. While enjoying alongside the children the immediate natural scenery, our aim is to nurture our community, partake in abundant conversations with our companions and build a healthy work life balance.

Currently, we are interested in the changes in the surroundings and the society that can stem from the conditions of the built environment. Instead of thinking of each building as
an independent entity, we propose to connect in a subtle way the people's activities with the exterior characteristics in the framework of the society that is part of this region. Although each plot of the town is of private property, we believe that it is more important to emphasize the public sphere rather than these private privileges.

In a society where people are connected loosely, instead of closing up the spaces, we
seek new spaces which are connected in a fluid way. While experiencing the area's real daily life, we strive to connect a scenery that goes hand in hand with emotion with the next generation.

>> ligth.jp

Neighborhood Facility

Location
Tokyo, Japan
Use
Office
Site area
250m²
Built area
83.25m²
Total floor area
250.59m²
Floors
3F

Exterior finish
Reinforced concrete, Acrylic silicone resin coating material
Interior finish
Exposed concrete
Project architect
Seiki Murashige
Photographer
Joaquin Mosquera [idearch]

MH Office

MH 오피스

GENDAI SEKKEI CO.,LTD.

Neighborhood Facility

SITE PLAN

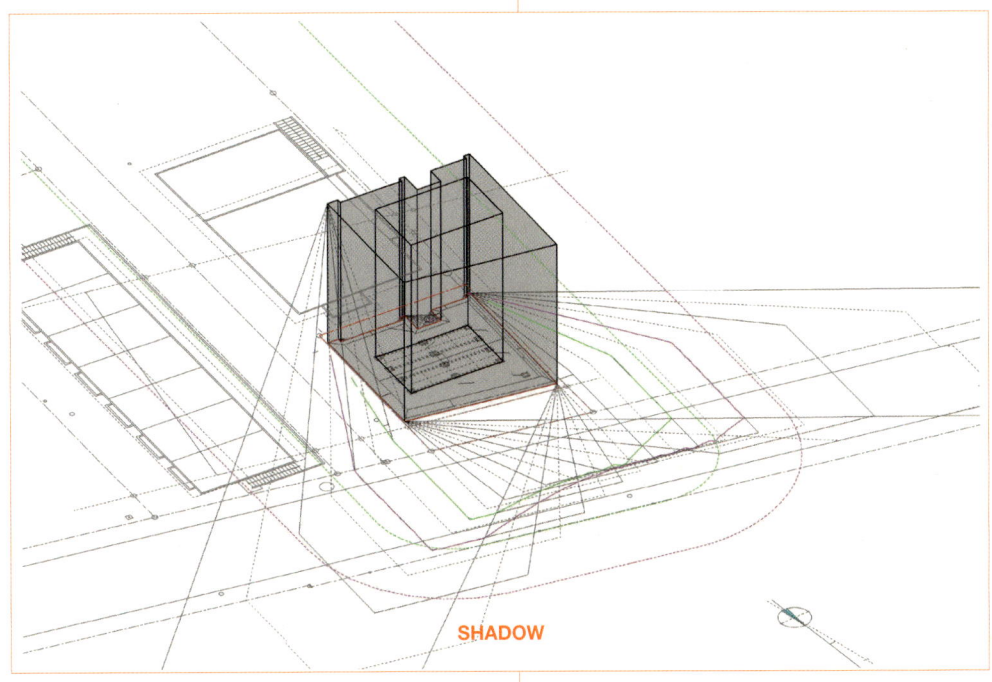

SHADOW

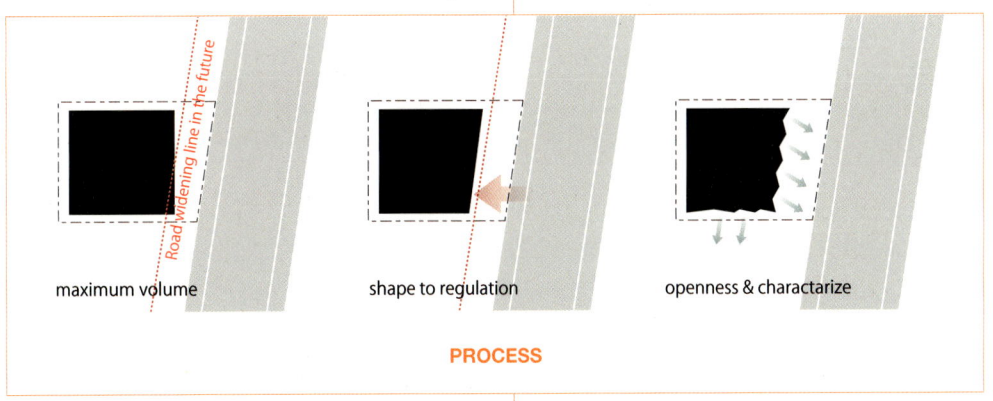

maximum volume   shape to regulation   openness & charactarize

Road widening line in the future

PROCESS

An office building that expresses the multifaceted character of a company with a folded plate-shaped façade

It is an office building for "Holdings," controlled by a logistics company based in Mitaka, west of Tokyo. The site facing the metropolitan highway runs north to south. After being divided in consideration of its actual use in the future, it will be phased out, and the apartments will be expanded in stages according to the first phase plan.

Holding is a system that spans various companies, so the architecture, which is the face of the company, is expected to be not uniform but multifaceted. At the time of planning, we thought a building that embodied the certainty of work and the company's rigorous rules originating in the logistics and transportation industries would be appropriate.

Due to the nature of the building, it is not advisable to bury it in the environment, but it is not desirable for its existence to be too prominent. Therefore, to satisfy such conditions as to privacy and openness, consideration for noise and vibrations, avoid the inaccessible environment of the outside and make the building intimate, each of the four turns of the building has different characters.

On the north side, which is highly visible from the outside, the openings and walls are repeated regularly. On the east side, the openings and walls are repeated with different spans in anticipation of an extended elevation plan at the time of expansion. In both cases, the folded plate-shaped outer wall is brought into the interior. On the south and west sides, planters and sliding windows will appear depending on the adjacent houses and apartments.

Neighborhood Facility

Based on the requirement to install solar power generation equipment on three floors plus a ceiling with absolute height restrictions, we kept the floor height as low as possible and pushed equipment and wiring pipes, which tend to be dark, to the ceiling. We set each floor's ceiling height to be slightly low at 2,300mm. However, the full-height windows, which are alternately arranged with the folded-plate-shaped walls, allow the line of sight to pass in a vertical direction. The interior has moderate privacy and a large expanse without feeling low or cramped.

With a simple but powerful stance, the architecture's appearance changed a lot with the installation of furniture; our goal is to build a building that could stand the test of time as the home of a deeply rooted company in the region.

절판 모양의 외벽으로 기업의 다면성을 표현한 오피스 빌딩

여기는 도쿄 서쪽에 위치한 미타카에 본사를 둔 물류회사가 관리하는 "지주회사" 사무실 빌딩이다. 대도시 고속도로를 마주하고 있는 부지는 남북으로 뻗어 있다. 향후 실 사용을 고려해 구획을 나눠 단계적으로 철거한 뒤 1단계 계획에 따라 단계적으로 아파트를 확대할 예정이다.

지주회사는 다양한 기업들이 운영하는 제도여서 기업의 얼굴인 건축 양식이 획일적이기 보다는 다면적일 것으로 예상된다. 기획 당시 물류와 운송 산업에 바탕을 둔 기업의 엄격한 규칙과 업무의 확실성을 구체화한 빌딩이 적절하다고 생각했다.

빌딩 특성상 주변환경에 묻히는 것은 적절치 않지만 존재감이 지나치게 부각되는 것 또한 바람직하지 않았다. 따라서 사생활과 개방성, 소음과 진동에 대한 고려, 외부로부터 접근이

DEVELOPMENT

불가능한 환경은 피하고 친밀한 빌딩을 만들기 위해 건물의 4개 방향마다 특성을 다르게 했다.

바깥에서 잘 보이는 북쪽에는 출입구와 벽이 규칙적으로 반복된다. 동쪽에서는 확장 시 입면 부분의 확장을 예상하여 출입구와 벽이 다른 간격으로 반복한다. 두 경우 모두 절판 모양의 외벽이 내부로 들어간다. 남쪽과 서쪽에는 인접한 주택과 아파트에 따라 화분과 슬라이딩 창이 보인다.

태양광 발전설비를 3층에 설치하고 절대 높이 제한이 있는 천장을 추가로 설치해야 한다는 요구조건을 바탕으로, 바닥 높이를 최대한 낮게 유지하고 어두워 보이기 쉬운 장비와 배선관을 천장으로 밀어냈다. 각 층의 천장 높이는 2,300mm로 약간 낮게 설정했다. 그러나 절판 모양의 벽과 번갈아 배치된 통창으로 시선이 수직 방향으로 통과하도록 했다. 실내는 낮은 느낌이나 비좁은 느낌 없이 적당한 프라이버시를 갖추고 공간은 넓다.

단순하지만 강렬한 모양으로 가구를 배치해 건축 양식의 모습을 많이 바꿨다. 본 프로젝트의 목표는 지역에 깊이 뿌리 내린 회사의 본거지로서 시간도 비껴간 빌딩을 짓는 것이다.

Neighborhood Facility

1 ENTRANCE
2 CORRIDOR
3 OFFICE
4 STORAGE
5 EQUIPMENT
6 PS/EPS
7 STAIR
8 RESTROOM
9 KITCHEN
10 BAR
11 ARCHIVE
12 ROOF

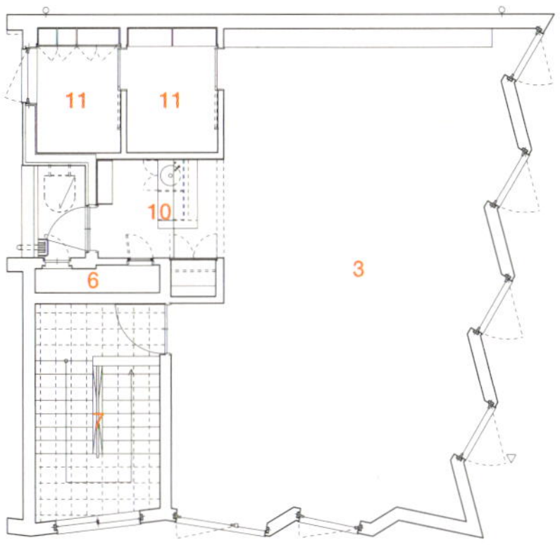

3RD FLOOR PLAN

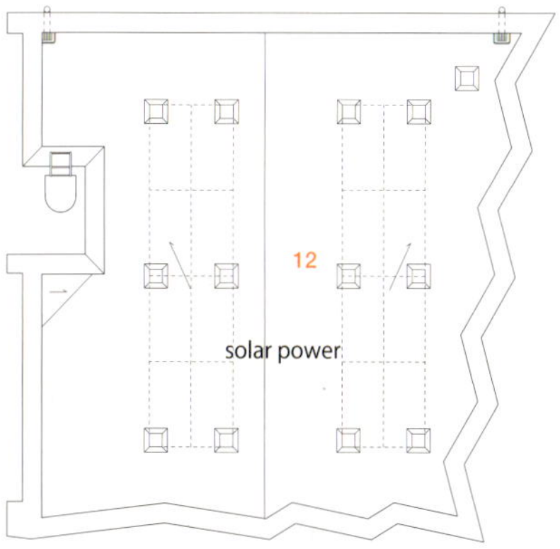

ROOF FLOOR PLAN

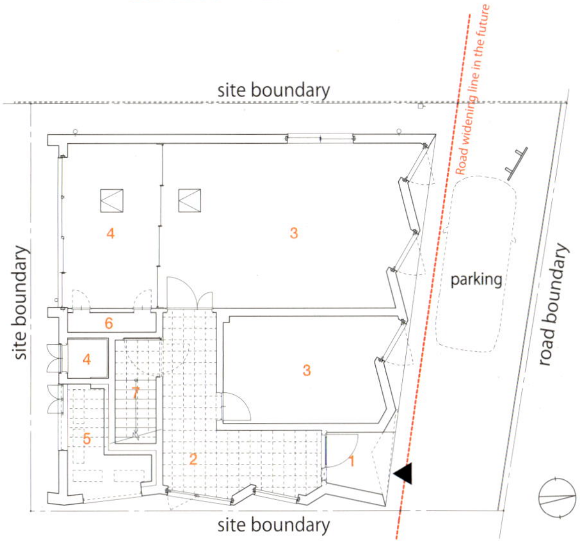

1ST FLOOR PLAN

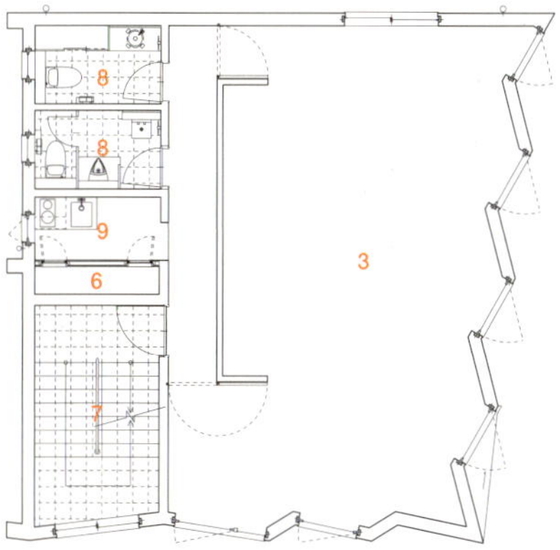

2ND FLOOR PLAN

Neighborhood Facility

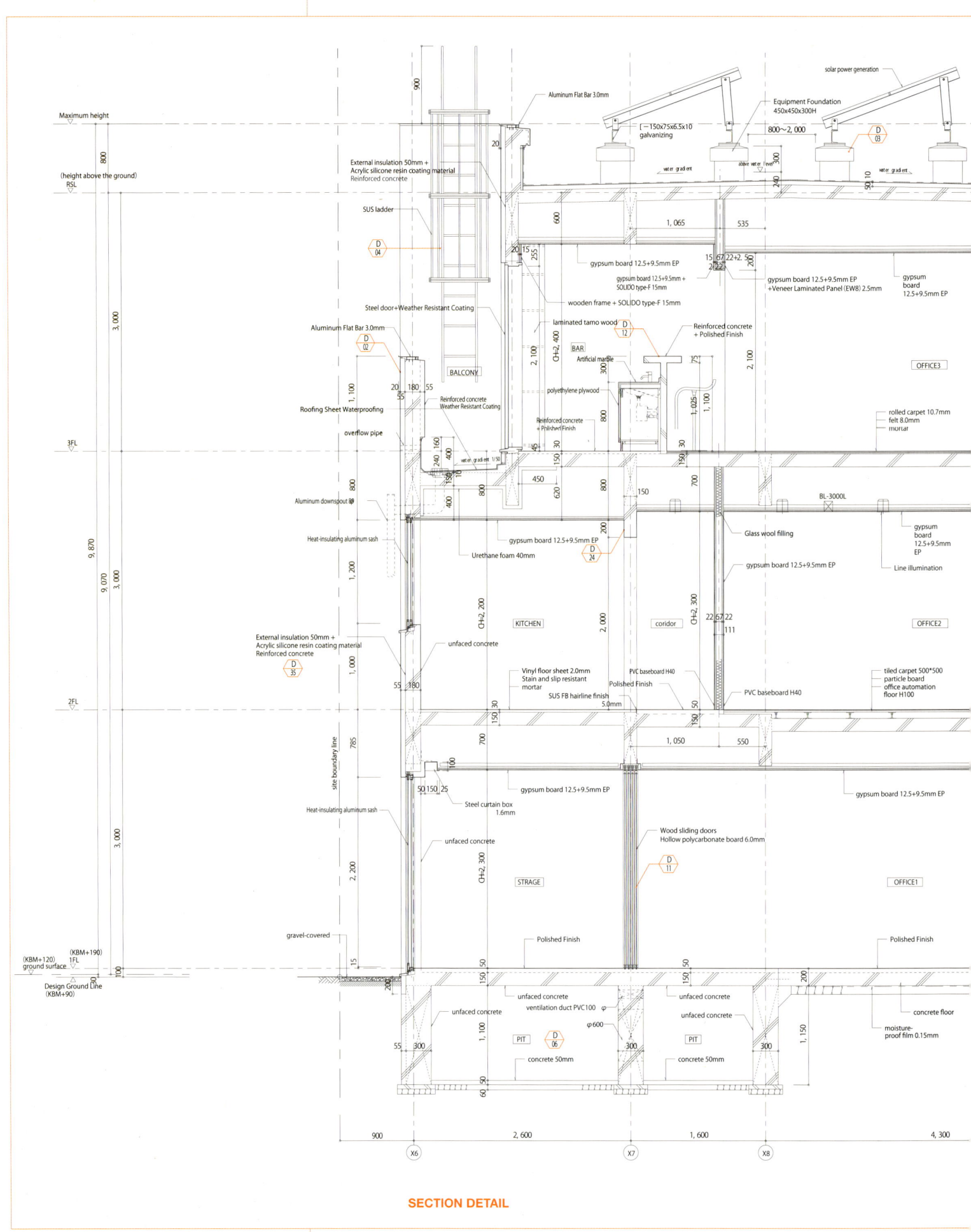

SECTION DETAIL

Neighborhood Facility

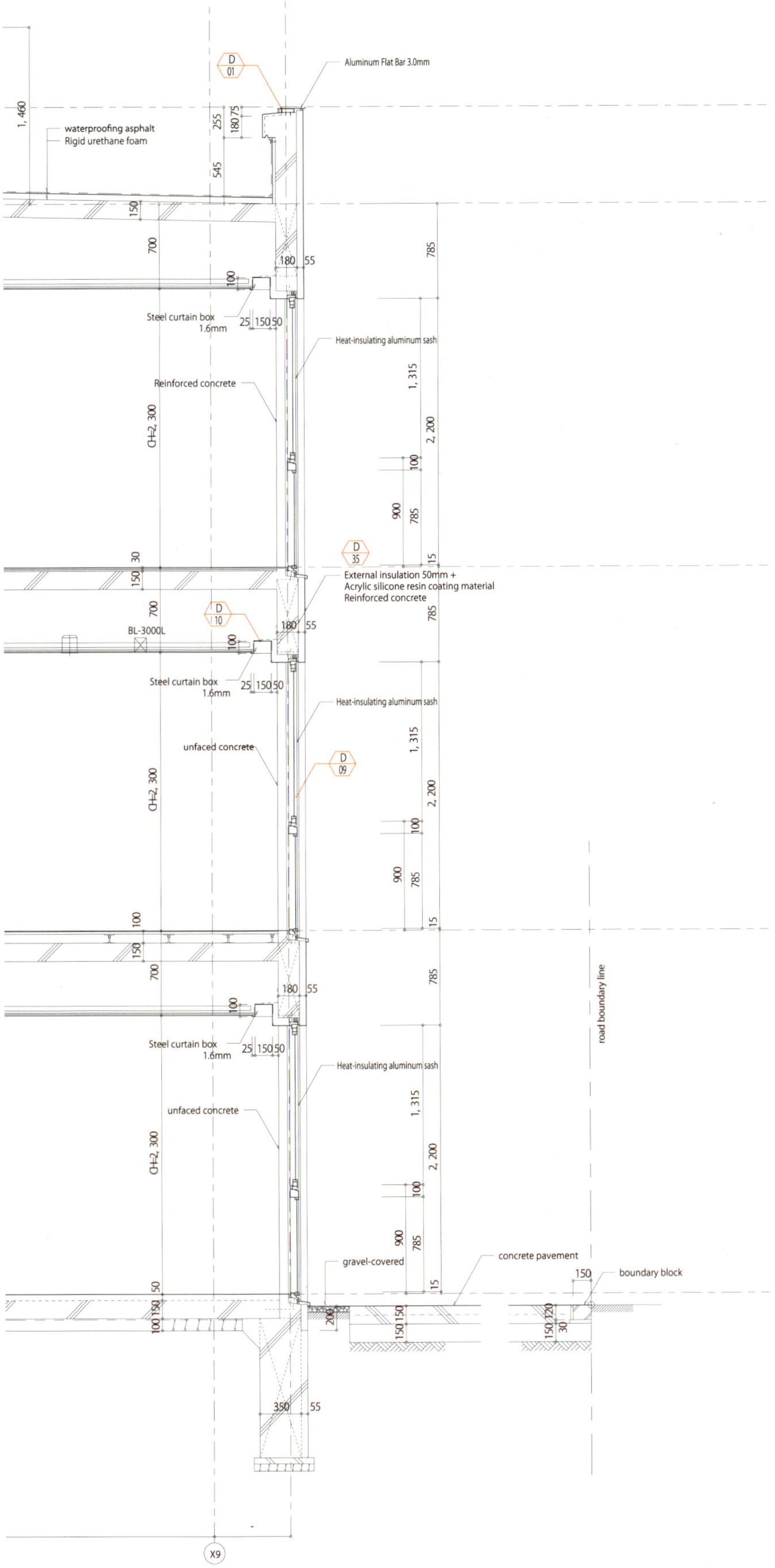
D 01
Aluminum Flat Bar 3.0mm
1, 460
waterproofing asphalt
Rigid urethane foam
255
180 75
545
150
700
180 55
785
100
Steel curtain box
1.6mm
25 150 50
Heat-insulating aluminum sash
Reinforced concrete
1, 315
CH=2, 300
2, 200
100
900
785
30
D 35
15
150
External insulation 50mm +
Acrylic silicone resin coating material
Reinforced concrete
700
D 10
180 55
785
BL-3000L
100
Steel curtain box
1.6mm
25 150 50
Heat-insulating aluminum sash
unfaced concrete
1, 315
D 09
CH=2, 300
2, 200
100
900
785
100
15
150
785
700
180 55
100
Steel curtain box
1.6mm
25 150 50
road boundary line
Heat-insulating aluminum sash
unfaced concrete
1, 315
CH=2, 300
2, 200
100
900
785
gravel-covered
concrete pavement
150
boundary block
50
15
100 150
200
150 150
150 120
30
350
55
X9

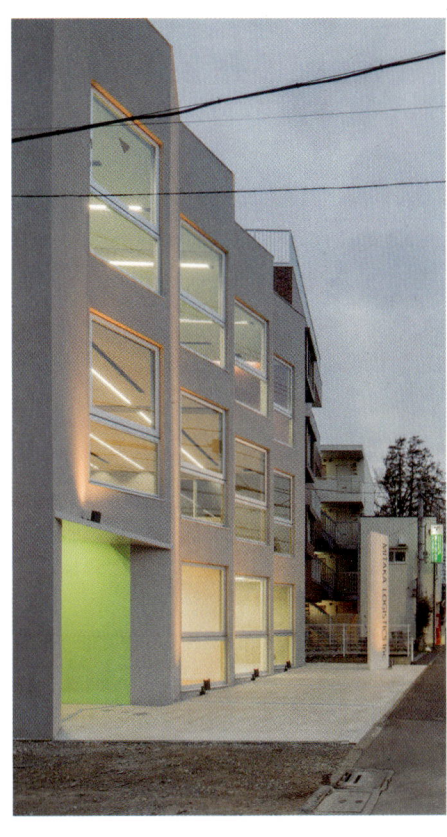

Neighborhood Facility

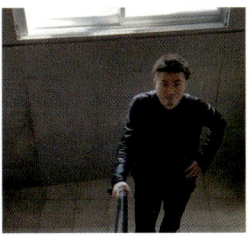

Seiki Murashige / GENDAI SEKKEI CO.,LTD.

A design office based in Yamaguchi, Hiroshima, and Tokyo with the basic philosophy of "Contributing to society through polite work. People who do polite work grow, and the growth of people becomes the development of the company."
The word "design" can be used to describe a wide variety of design "work". What you see in the public eye is just one of them.
New construction, renovation, partial repair, demolition, investigation, restoration, periodic reporting, seismic diagnosis, seismic reinforcement, etc. We believe that there is always something to learn in every design job, and we carry out our design activities every day.

>> gendaisekkei.co.jp

Location
Vadodara, Gujarat, India
Use
Office
Site area
125.4m² (1,350ft²)
Built area
83.6m² (900ft²)
Total floor area
418m² (4,500ft²)

Project architect
Manoj Patel
Photographs
MK Gandhi Studio

Architecture and interior design studio office

건축&인테리어 디자인 스튜디오 사무실

Manoj Patel Design Studio

Manoj Patel Design studio's new workspace spreads over plot area of 1,350 sqft. surrounded by the urban fabric, revolves around the play of volumetric masses juxtapose with colorful graphics. To create an experience of material palettes, touch, feel and understand them in detail was the main idea to create a CO-Working studio space. On entering, the studio greets with bold adaptive artworks on the entry gateway from clay tiles infused with stone work and colors. A studio's facelift seems to be metaphor with natural habitat around. Floral variations seem to be sculpturesque line between studio space and art installations. Connecting amphitheater inside has a bespoke feature wall with minimal ceiling that sets to enjoy the celebrations of making an avenue connecting, the design community and becoming a place for experiments, learning and sharing ideas together.

The visual identity uses bold polychromatic shades with voluminous spaces to convey a vibrant visual of youthfulness and story tales inside the workspace.

Giant translucent metallic door welcomes within the main office space. An elongated spatial arrangement of reception and an informal waiting area signifies an open feature of welcome lobbies. The backdrop mural's aspiration was to craft clay into delicate wall hangings that are based on curves and creates curiosity from around when passersby visit. Lift wall is embodied with the clay tile and mosaics tiles cluster highlights the variants inside the space.

A series of intensive design layouts, gave rise to create an entertaining triple height corner of pockets with activities, 3D metal murals and clay modules on the perpendicular wall. One can visually feel this connectivity through barrier free openings. Horizontal and vertical curving geometry of colors meticulously introduce a new form of material palette collection on wall. A new

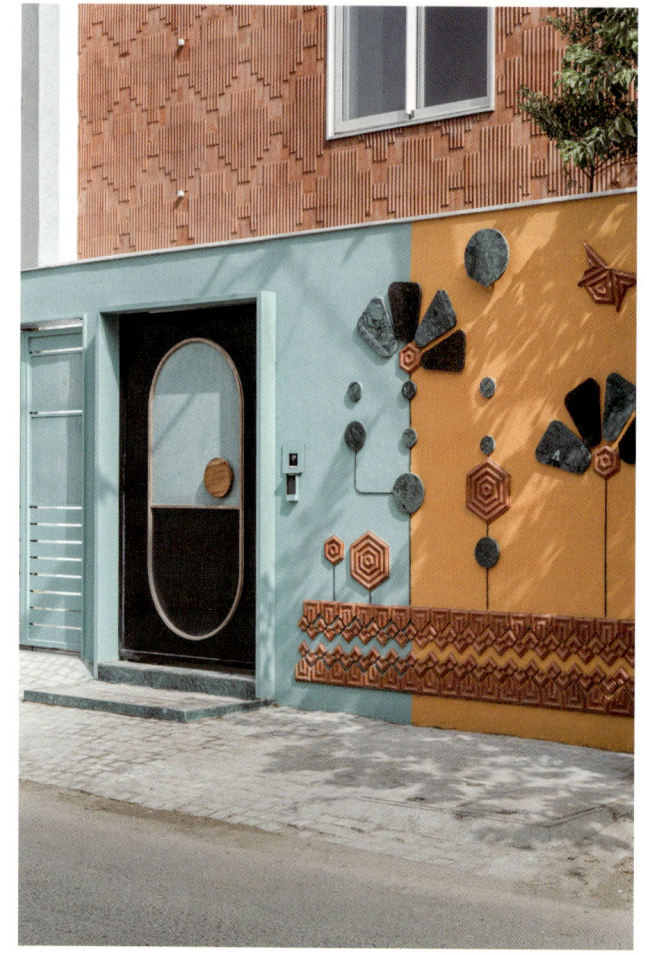

*On entering, the studio greets with bold adaptive art-works on the entry gateway from clay tiles infused with stone work and colors. A studio's facelift seems to be metaphor with natural habitat around. Floral variations seem to be sculpturesque line between studio space and art installations.

*CLAY TILE: Facade explores another surface treatment of studio's hand crafted material details. Strips of vernacular material are oriented to form free florals forms that change accordingly.
Overlapping clay profiles with subtle blue color and balcony canopies, pops out for the facade volume.

COLOR

ideology is taken into consideration while designing the triple height mural wall, which can evolved over the times with new ideas exploring fabric for new generations.
Façade explores another surface treatment of studio's hand crafted material details. Strips of vernacular material are oriented to form free florals forms that change accordingly. Overlapping them with subtle blue color and balcony canopies, pops out for the façade volume.
The convertible Principal Architect's cabin into a conference area collaborates with a sit out space to have informal conversations and meetings as part of different experience. Spaces are arranged as pockets of open and enclosed around the central waiting lounge on the first floor.
This waiting lounge is based on idea of creating a place with unique combinations of organic shapes, neutral colors and curved graphics on wall.
The triple height building connects the spaces via different openings, which mainly acts more as punchers for more connectivity. On the above floors for the staffs work desk, a thoughtful intervention has been added. A bridge with double height connections within the above 2 floors, to connect with fellow teammates and gets to learn or pass on learning easily without any barriers. A series of different visual working pockets turns out to be combination of work, interaction and relaxation for the staff as well as the clients who visit.
While travelling from one floor to another there are tile backdrops designed which acts as a engaging storytelling concept inside the studio.
Studio is always willing to experiment with new things and continuously develop them into executions. For this the office came up with an in-house material studio where the manufacturers can directly interact with the project leads and integrate collaborative

137

processes. Considering new perspectives and technological upliftments in the same sector, digitalization has also been added into this space. Cantilever balcony projections ensure safety and generate curiosity of communications with one another in the open air.

Colorful materials display gradient effect on subtle walls. One of the accent walls exhibits possibilities on materials can be altered as per applications and can be used numerous times as per designer's thoughts.

Co-Working space divides functions based on public and private intervention in the building on above floors. Thinking out of the box, studio incorporated penthouse on the last 2 floors that serves to be relaxing zone for the staff along side introducing the clients to the technologies and layout orientations fitted inside. Barrier free studio living encompasses the open minded spaces yet catering to private needs. An informal high chair seat out on the terrace garden has green and self sufficient biophilic designs that breathe natural light and ventilations. Play of mood setting lights and music reestablishes connect with working in nature for drawing inspirations and progressive character of mind.

마노즈 파텔(Manoj Patel) 디자인 스튜디오의 새로운 작업 공간은 도시조직(urban fabric)으로 둘러싸인 면적 1350평방피트의 건물로, 화려한 그래픽과 부피가 큰 구조물이 그 특징이다. 재료적 팔레트를 경험하게 하고자 팔레트를 자세히 만지고 느끼고 이해하는 것이 공동작업 스튜디오 공간 조성의 주요 아이디어였다.

스튜디오에 들어서면 석조물과 다양한 색상의 점토 타일이 입구에서부터 대담한 적응형 예술작품으로 인사를 건넨다. 스튜디오의 변신은 주변 자연 서식지의 또 다른 메타포어이다. 꽃의 변주는 스튜디오 공간과 예술 설치물의 조각 같은 선처럼 보인다. 내부로 연결되는 계단식 공간은 최소한의 천장만 갖춘 맞춤형 기능 벽이 있어 길가와 이어지고, 디자인 커뮤니티를 조성하며 실험, 학습 및 아이디어를 함께 공유하는 장소가 되어 이벤트를 즐길 수 있다.
시각적 특징으로는 대담한 다색의 음영과 볼륨감 있는 공간을 사용하여 작업 공간 내부의 젊음과 스토리텔링의 생생함을 시각적으로 전달한다.

A new ideology is taken into consideration while designing the triple height mural wall, which can evolved over the times with new ideas exploring fabric for new generations.

DESIGN THE TRIPLE HEIGHT MURAL WALL

거대한 반투명 메탈 문이 주 사무 공간에서 사람들을 환영한다. 길쭉한 리셉션 공간 배치와 격의 없는 대기 공간은 환영의 분위기가 있는 로비의 열린 특징을 나타낸다. 배경에 벽화를 둔 이유는 곡선을 바탕으로 한 섬세한 점토 벽걸이를 통해 이곳을 방문한 외부인들이 주변에 호기심을 갖도록 하는 목표에서 비롯되었다. 리프트 월은 점토 타일로 구체화해 모자이크 타일이 공간 내부의 변화를 강조한다.

집중적인 디자인 배치를 통해 재미있는 3층 높이의 포켓 코너를 만들었는데, 수직 벽에서 다양한 활동을 하거나 벽화 및 점토 모듈을 걸 수 있다. 막힘 없는 출입구를 통해 이러한 연결성을 시각적으로 느낄 수 있다. 색상은 수평 및 수직 등 기하학적 곡선 형태로 배치함으로써 벽면에 새로운 형태의 재료 컬렉션을 세심하게 도입한다. 새로운 사상을 고려하는 한편, 시간이 지남에 따라 새로운 세대를 위한 조직을 탐구하는 새로운 아이디어와 함께 진화하는 3중 높이의 벽화 벽면을 디자인했다.

정면부는 스튜디오에서 수작업으로 제작한 소재 디테일의 또 다른 표면 처리를 탐구한다. 현지 소재로 만든 스트립으로 적절히 변화하는 자유로운 꽃 모양을 만들었다. 은은한 파란색과 발코니 캐노피가 겹쳐져 정면부의

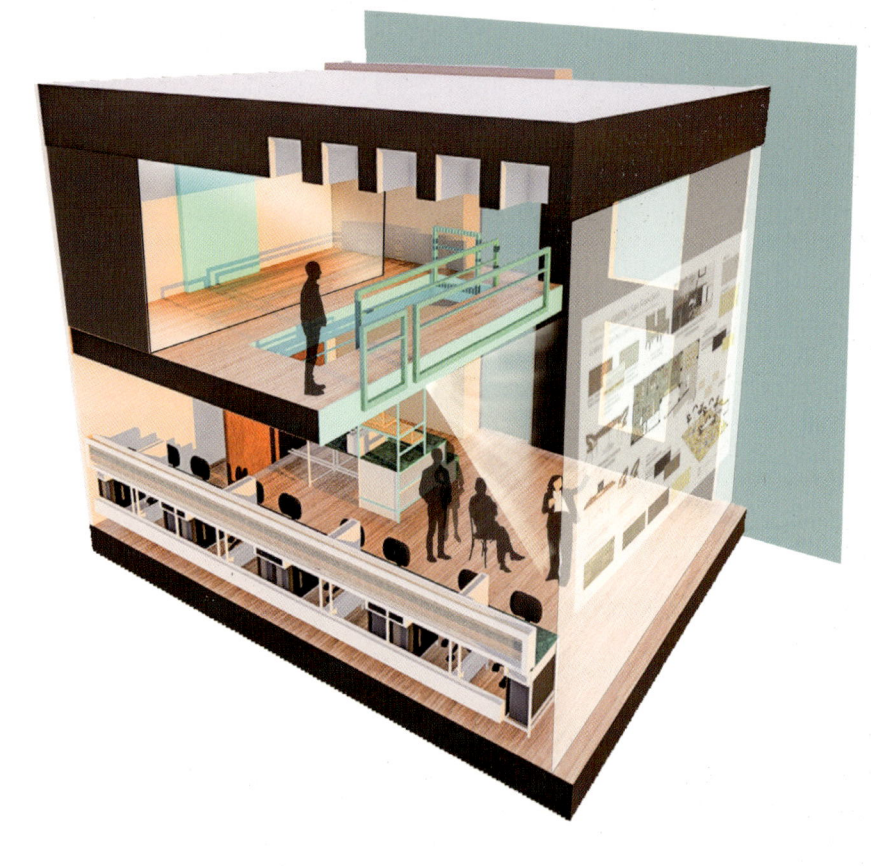

Neighborhood Facility

볼륨감이 돋보인다.
대표건축가 사무실은 회의 공간으로 전환되어 휴식 공간과 편안한 대화의 장이라는 또 다른 용도로서 사용이 가능하다. 공간은 1층 중앙 대기 라운지를 중심으로 개방형 및 폐쇄형 포켓으로 배치된다. 이 대기 라운지는 벽에 유기적인 형태와 중성적인 색상, 곡선화된 그래픽이 독특한 조화를 이루는 공간을 조성하자는 아이디어를 바탕으로 만들었다.

3층 빌딩은 서로 다른 출입구를 통해 공간이 연결되는데, 이로써 연결성이 더 많이 부각된다. 위층의 직원 업무용 책상에는 사려 깊은 개입이 추가되었다.
상부 2개 층은 이중 다리를 통해 팀원들과 연결되어 장애물 없이 쉽게 소통할 수 있다. 일련의 다양한 시각 작업용 포켓은 직원과 방문 고객을 위한 작업, 상호작용 및 휴식이 모두 이루어질 수 있는 공간이다. 스튜디오 안에는 한 층에서 다른 층으로 이동하는 동안 매력적인 스토리텔링 개념으로 작용하는 타일 배경을 디자인했다.

스튜디오는 기꺼이 새로운 것을 실험하고 실행을 통해 새로운 것들을 지속적으로 발전시켜 나간다. 이를 위해 사무실은 제조업체가 프로젝트 리더들과 직접 상호작용하고 협업 단계를 통합할 수 있는 사내 머티리얼 스튜디오를 마련했다. 동일한 분야에서의 새로운 관점과 기술적 향상을 고려하여 디지털화 또한 이 공간에 추가했다. 캔틸레버 발코니 돌출부는 안전을 보장하는 동시에 옥외 소통에 대한 호기심을 유발한다. 다채로운 색상의 소재들은 은은한 벽면에 그라데이션 효과를 발휘한다. 어느 액센트 벽은 소재 가능성을 보여주며 용도에 따라 교체할 수 있고 디자이너의 생각에 따라 여러 번 사용할 수도 있다.

공동작업 공간은 상부 층에서 공공 및 사적인 개입이 얼마나 있는지를 기준으로 그 기능을 나눈다. 스튜디오는 고정관념에서 벗어나 마지막 2개 층에 펜트하우스를 통합하여 고객에게는 빌딩 내부에 설치된 기술과 레이아웃을 소개하는 동시에 직원들을 위한

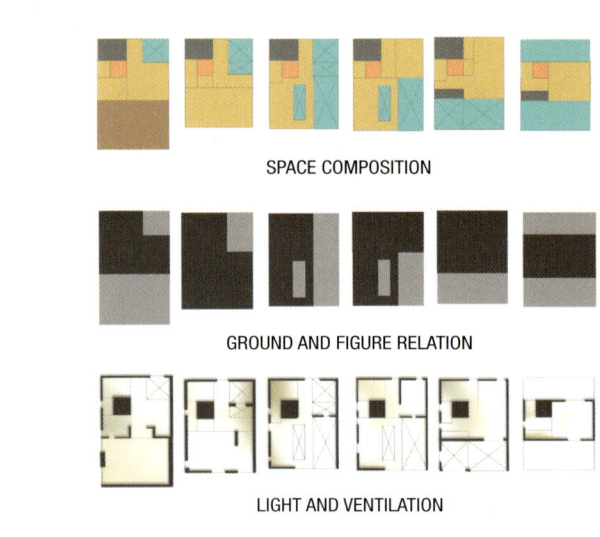

CONCEPT PLANNING

SPACE COMPOSITION

GROUND AND FIGURE RELATION

LIGHT AND VENTILATION

VOLUMETRIC SECTION

휴식 공간을 만들었다. 막힘이 없는 스튜디오 생활은 열린 마음의 공간을 포함하지만 개인적인 필요 또한 충족시킨다. 테라스 정원에 있는 캐주얼한 높은 의자는 자연광을 받고 환기를 통해 호흡할 수 있는 녹색의 자급자족형 생물 친화적 디자인을 띤다. 무드등을 켜고 음악을 틀면 자연 속에서 업무를 보는 상태가 조성되어 영감과 기발한 생각을 이끌어 낸다.

SECTION B-B'

SECTION

Neighborhood Facility

THIRD FLOOR LAYOUT　　　FOURTH FLOOR LAYOUT　　　FIFTH FLOOR LAYOUT

GROUND FLOOR LAYOUT　　　FIRST FLOOR LAYOUT　　　SECOND FLOOR LAYOUT

PLAN

1　3　6

N

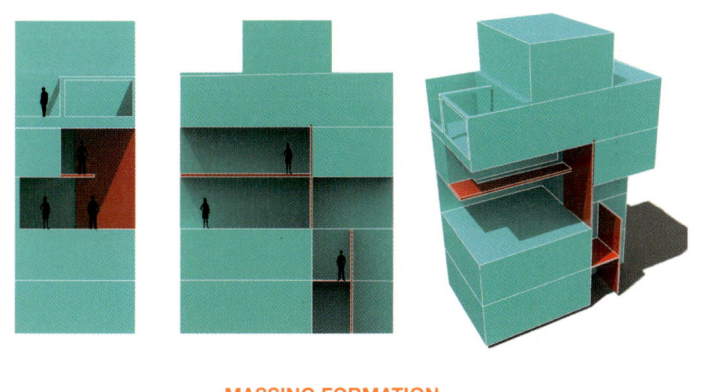

MASSING FORMATION

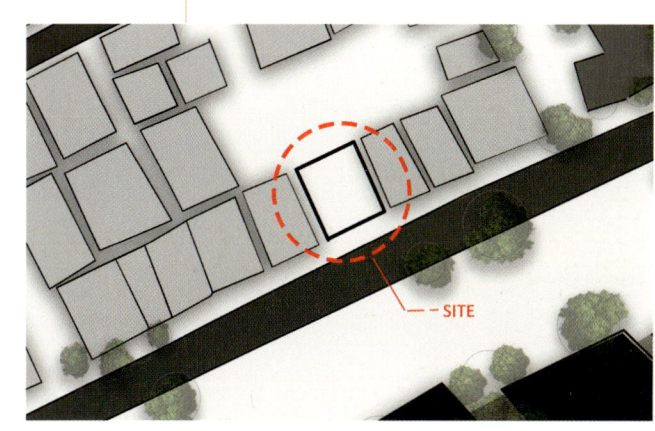

SITE

143

SECTION A-A'

1 3 6

SECTION

Neighborhood Facility

Manoj Patel Design Studio

The firm Manoj Patel Design Studio is nestled in Vadodara. Manoj Patel is principal architect at Manoj Patel Design Studio. Manoj Patel have B.Arch from A.P.I.E.D (Gujarat, India) & M.Arch in Sustainable Architecture from CEPT University (Ahmedabad, Gujarat, India).

The aim of the firm is keen about climate responsive architecture and reviving old traditional practices. They believe in designing projects of all scales where they can give our clients new way of living into the spaces or built environment. Studio's work majorly focuses on sustainable building designs by exploring recyclable materials in built contemporary forms. The firm encourage local craftsmen in our project where they get encouraged and the Indian art is revived.

>> www.manojpateldesignstudio.com

Redefining the idea of combining work and home in a rather novel way. The office space encourages healthy living where staff can feel like home, rest-relax and work for long hours and get meals cooked in-house. Clients who travel long distances can take a nap here. The vibe ranges from casual and cafe-like, to more formal leading to better efficiency.

Connecting amphitheater inside has a bespoke feature wall with minimal ceiling that sets to enjoy the celebrations of making an avenue connecting, the design community an becoming a place for experiments, learning and sharing ideas together.

Neighborhood Facility

NINE CUBE

나인 큐브

(주)서원건축사사무소
SEO WON ARCHITECTS COOPERATION

Location
Busan, Republic of Korea
Use
Neighborhood facility (Office)
Site area
237.00m²
Built area
85.83m²
Total floor area
190.78m²
Floors
3F

Exterior finish
Stucco, Steel plate
Interior finish
Exposed brick, Plywood
Project architect
Cho Seo-young
Construction
SEO WON Architects Cooperation
Photographer
Yoon Joonhwan

Neighborhood Facility

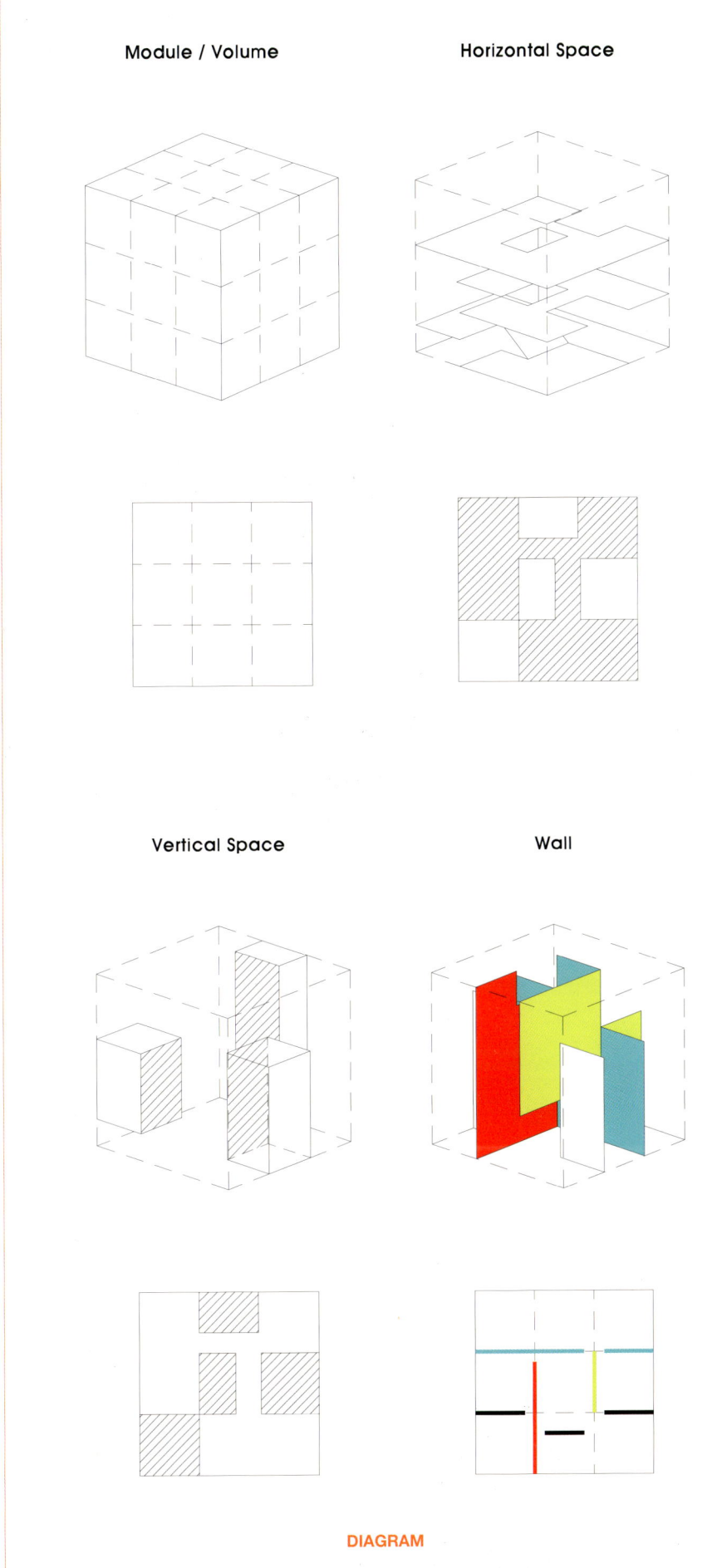

DIAGRAM

계획개념
쾌적한 업무환경과 내 외부 유기적 공간을 가진 사무실건축을 실현하고자, 공간감 구현과 유기적 업무환경을 위한 건축구축(構築, structure)이라는 개념으로 공간구성을 계획하였다.

VOID / SOLID
내부는 비움과 채움의 개념을 적용하여 각 층마다 열리고 닫히며 수직적으로 연결되고, 수평적으로 열리는 공간이 다시 이어지도록 내부공간 개념을 계획했다.
정적인 사무공간이 동적인 공간과 연계가 되도록 하였으며, 진입 1층은 외부에 개방할 수 있도록 하였다.

MODULE / VOLUME
기본적 구성은 3m×3m모듈을 3×3열로 배열하여 총 9개의 모듈이 9m×9m 정사각형으로 평면구성되며, 이 9m×9m 정사각형이 3개층의 입방체 볼륨(volume)으로 입체구성 되어있다. 모듈화 과정을 통해, 3m×3m 모듈이 총 27개로 9m×9m×9m 입방체가 된 것이다. 각 모듈에는 개별적인 목적과 경제적 합리성을 기준으로 하여 구축방식과 재료를 선정하였고, 각각의 성격을 부여하여 연속적이면서도 개별적인 특성을 갖게 하였다.

사무실 건축 프로젝트로서 합리적이고 경제적인 건축물을 실현하기 위하여, 모듈화된 철골건축으로 결정하였다. 이 때문에 재료자체가 가지고 있는 멋을 그대로 보여주고자 하였으며 다양한 공간적 힘이 있기를 바랐다. 오랜 마을 고촌(古村)에서 어색하지 않고 활달한 젊은 이웃이 되고자 한다.

Neighborhood Facility

Planning Concept
In office architecture, we have created a design that offers a pleasant work environment by realizing a spatial sense that ensures harmony between the building's interior and exterior

VOID / SOLID
The interior employs the concept of emptiness and fullness in every space. Each level opens and closes, connects vertically and opens horizontally while opening and closing on each floor. The static office spaces were intended to merge seamlessly with the dynamic areas. The entrance is open to the outside to provide easy access and a sense of openness.

MODULE / VOLUME
The fundamental structure has been arranged in a 3m x 3m module format, aligned so that nine modules would provide a flat composition of a 9m x 9m square, and three layers of the 9m x 9m square would create a three dimensional cube. The process of modular-based construction utilizes twenty-seven 3m x 3m modules, and each of the 9m × 9m × 9m cubes was built sequentially and individually with its own character by choosing construction methods and materials based on each individual purpose and economic rationality. Each module was designed to reveal its purpose and function to the outside world and respond to its environment.

A modularized steel structure was adopted, constructed, and completed to become a fair and cost-effective building in office architecture. Our aim was to display the beauty of the raw materials and create diverse spatial dynamics as a vibrant, youthful neighbor in a quaint village.

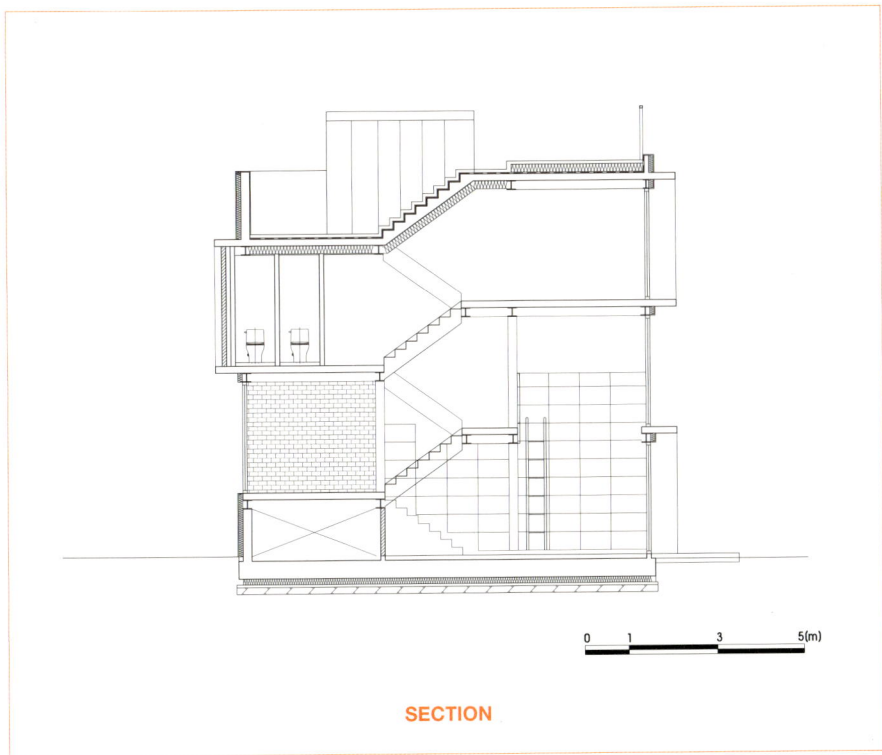

SECTION

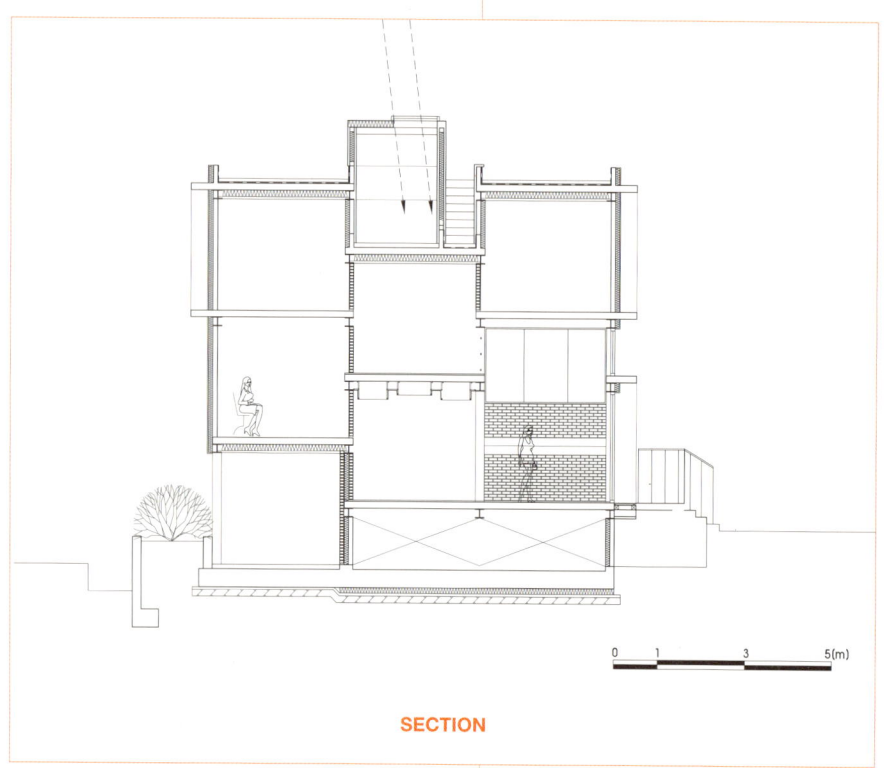

SECTION

Neighborhood Facility

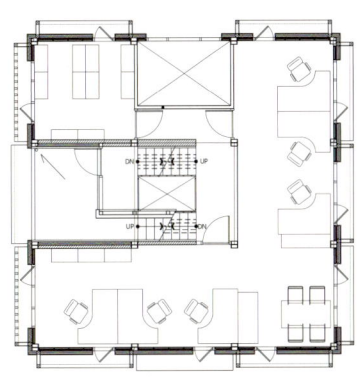

3RD FLOOR PLAN

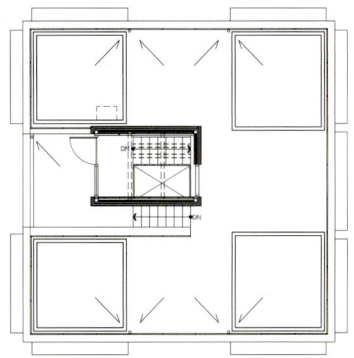

ROOF PLAN

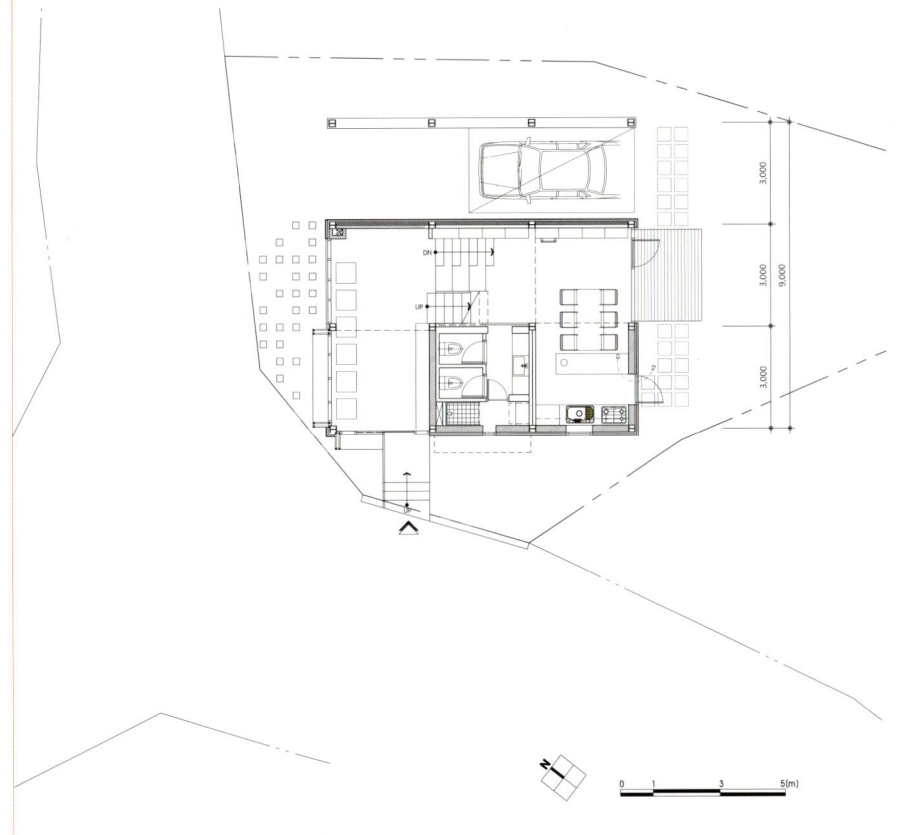

1ST FLOOR PLAN

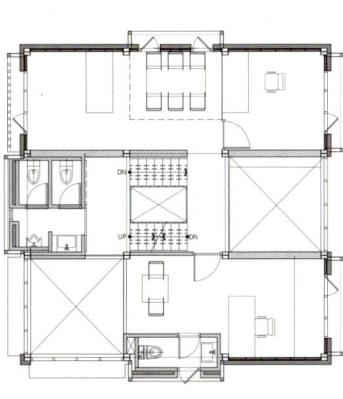

2ND FLOOR PLAN

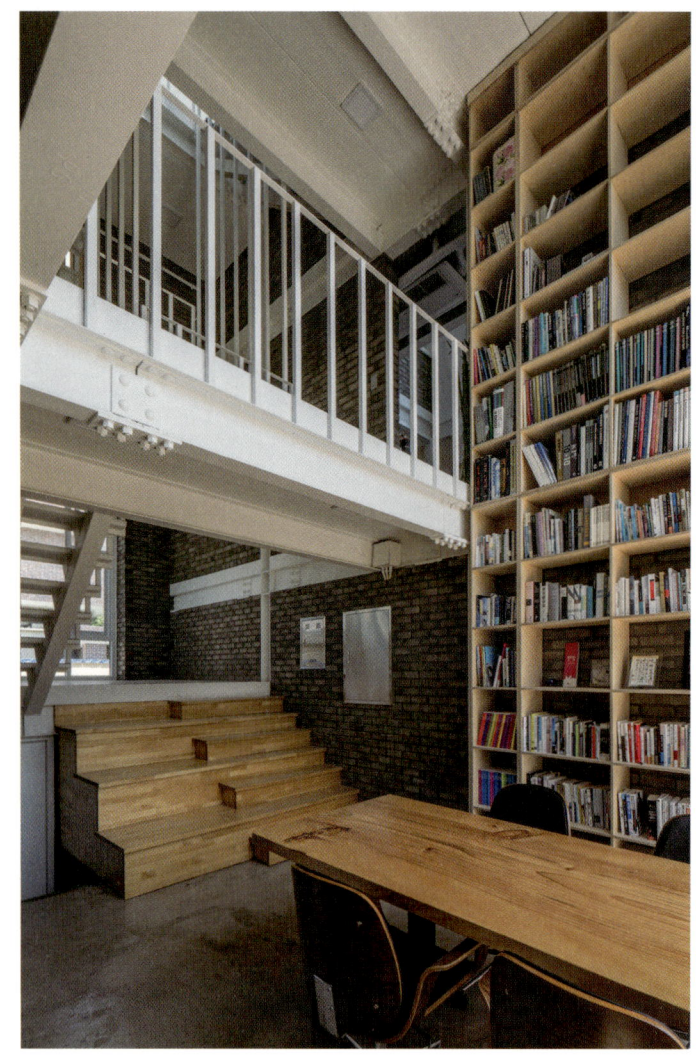
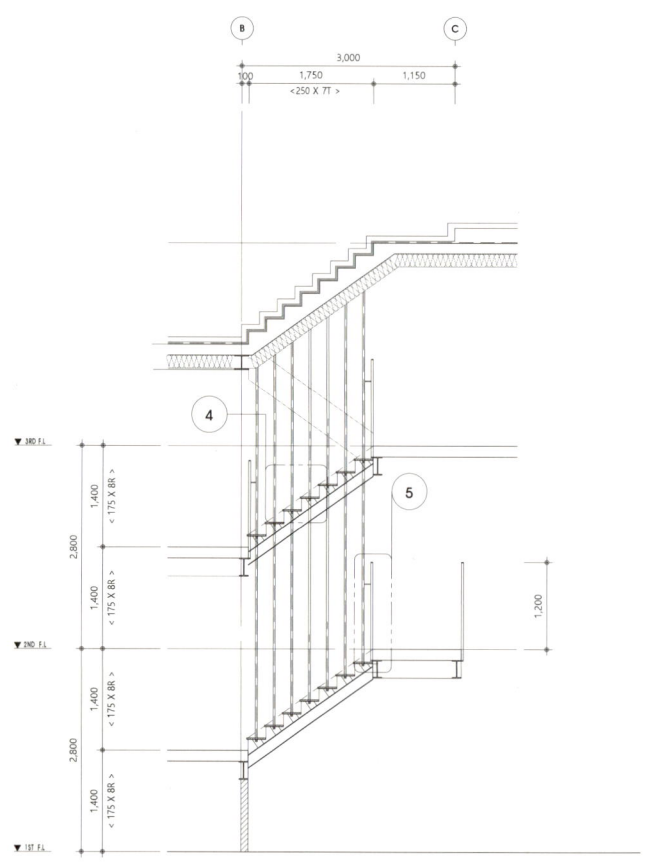

STAIR - SECTION

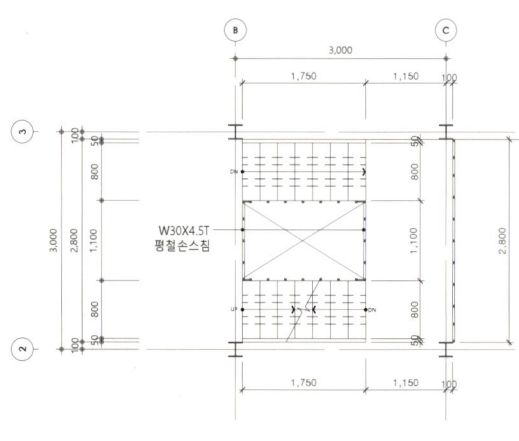

STAIR - PLAN

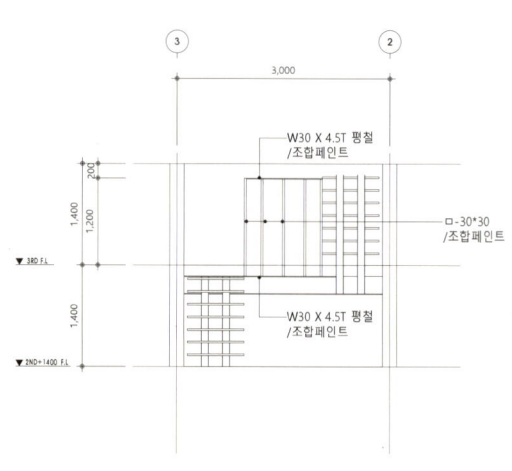

STAIR - ELEVATION

DETAIL

Neighborhood Facility

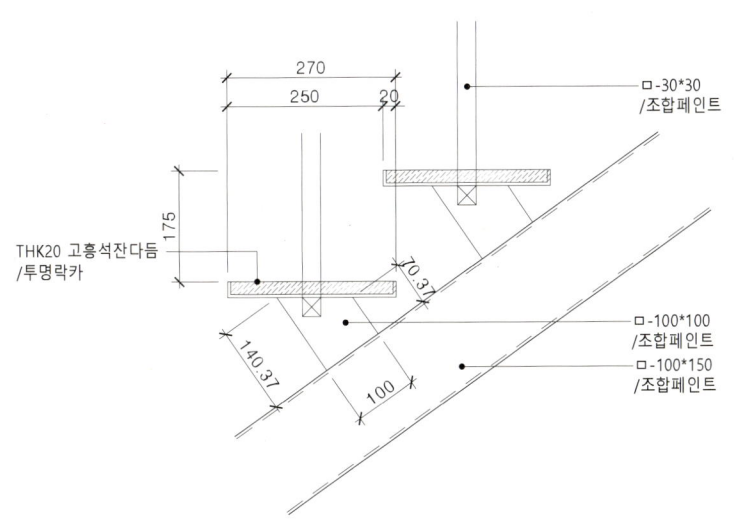

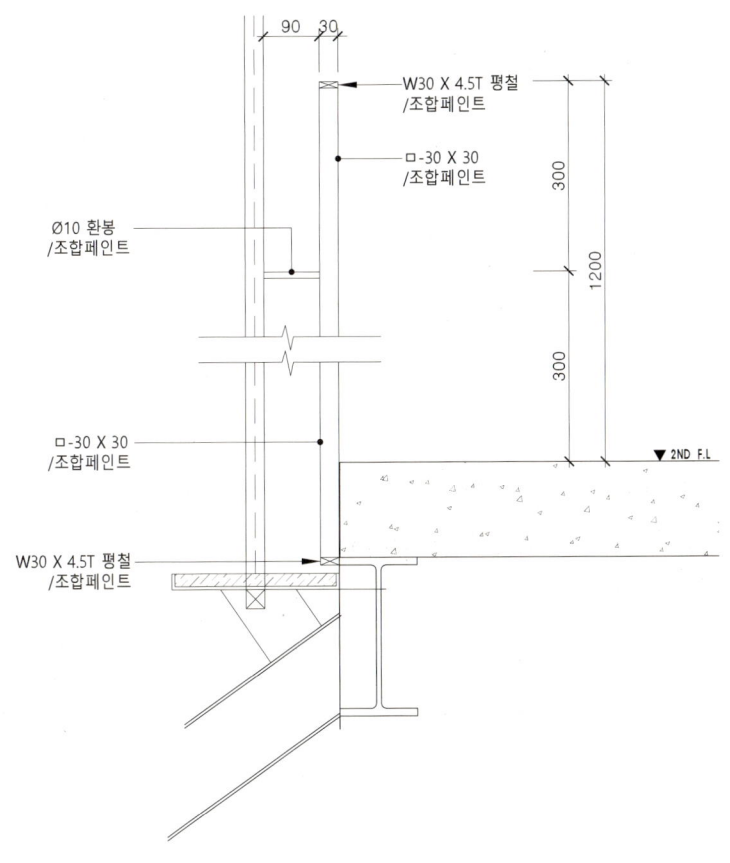

SECTION DETAIL

조서영 / ㈜서원건축사사무소
Cho Seo-young / SEO WON Architects Cooperation

동의대학교 건축공학과를 1991년에 졸업하고, 1998년에 건축사자격을 취득하였다. 2003년부터 2020년까지 동의대학교 및 동서대학교 건축학과에 출강 했으며, 2000년 3월에 서원건축사사무소를 개소하여 23년째 운영하고 있다.
(사)한국건축가협회 부산건축가회 회장(2020-2022)을 역임한 바 있고, 부산국제건축문화제 이사 및 부산시 교육청 공공건축심의위원으로 활동하고 있다.
주요작품은 글마루 작은도서관, 푸른솔경로당, 수영구노인복지관, 보수동주민센터, 해안경관 조명공간 역사의 디오라마, 금샘도서관 등으로 주로 공공건축물을 설계하였다.
수상경력으로는 2008, 2011년, 2012년 부산시장 표창장과 제3회 부산건축가회 신인건축가상, 2014 국토교통부 장관상 수상 등이 있다.

Neighborhood Facility

Location
Jongro-gu, Seoul, Republic of Korea
Use
Commercial, residential
Site area
149.50m^2
Built area
88.77m^2
Total floor area
197.88m^2
Floors
4F

Exterior finish
Exposed concrete, brick tile, wood
Interior finish
Exposed concrete, paint, wood
Project architect
Kim Younsoo
Construction
SOLHEIMconstruction.co
Photographer
Hwang Hyo Chel

JAEDONG PROJECT / HALF & HALF BUILDING

재동 프로젝트 / 반반건축

바운더리스 아키텍츠
BOUNDARIES ARCHITECTS

Neighborhood Facility

BEFORE

SITE CONTEXT

한옥지구와 도시조직의 경

재동프로젝트는 재동46-5 한옥과 도로를 만들며 잘려진 재동45-9 필지의 단층건물을 합필하여 개발하는 계획에서 시작하였다. 지구단위계획에 따라 12m, 4층까지 개발할 수 있는 구역의 경계의 대지이며, 한옥보전지구와 맞닿아 있었다. 최초의 계획은 ㄷ자 한옥 중 막다른 골목에 위치한 행간채를 남겨서 골목의 풍경을 유지하고 북촌로4길에서는 별동 증축을 하여 기존 한옥과 새건축물이 마당을 사이에 두고 마주보는 모습이었으나, 허가 중 한옥보전지구 지정이 되어 서울시와 종로구의 심의와 협의를 통해 한옥을 최대한 보존하며 증축을 하는 현재의 안으로 조정이 되었다.

남기고 덧붙이기

한옥을 최대한 보존하는 방향으로 계획이 전환되며 ㄷ자 한옥의 대청과 안채를 철거하여 ㄴ자 한옥으로 조정을 하게 되었으며, 당시의 조건 안에서 최대한의 연면적을 확보하기 위하여 남겨진 한옥에 새건축물을 결합해 건축하는 하이브리드 증축으로 진행이 되었다. 연결된 건축물이지만 개별 구조체로 분리가 되어있고, 외피로만 연결이 되어있다. 예전의 한옥에 있었던 마당의 공간을 유지하지만 한옥과 양옥이 공유하는 공간으로 변화하였다.

반반건축과 반사와 환영

외관에서 느끼기는 힘들지만 한옥과 새로운 건축은 서로 반반을 짤라서 붙인 듯이 내부공간을 구성하며 붙어있다. 짬짜면 혹은 반반세트와 같은 모습은 각각은 익숙한 것들의 조합이지만 매우 낯선 모습으로 흥미롭게 다가온다. 거기에 더불어 필로티 하부의 미러바리솔에 의한 반사된 기와지붕와 뒤집힌 풍경은 다른 세계로의 통로와 같이 느껴지며, 마당을 바라보는 창호에 반사된 한옥의 모습은 사라진 한옥의 환영처럼 과거와 현재를 중첩하여 느끼도록 만든다.

한옥 위 양옥

12m 높이 제한이 있는 상태에서 한옥을 유지하며, 요구하는 프로그램과 면적을 만족시키기 위해서는 일반적인 층의 개념을 가진 단면으로는 어려움이 있었다. 한옥의 지붕높이에 의해 반층씩 차이가 나는 스킵플로어 형태의 단면계획은 자연스럽게 대안이 되었으며 아주 작은 공간이라도 확보하기 위해 퍼즐을 끼워맞추는 단면과 공간의 형태가 제안되었다.

A Boundary between a Hanok District and the Urban Fabric

The Jaedong project started with a plan to develop the hanok at 46-5 Jaedong and combine it with the one-story structure at 45-9 Jaedong that had been cut off during road construction. The site is on the edge of a zone where the district unit plan permits developments of up to twelve meters or four stories high, and it is directly adjacent to the Hanok Preservation District. The original plan was to preserve the scenery of the alley by retaining a wing of the ㄷ-shaped hanok at the dead end and to construct an extra

building on the side of Bukchon-ro 4-gil so that the old hanok and a new structure would face each other with a courtyard in between. However, because the hanok was designated as a part of the Hanok Preservation District during the permission process, deliberation and consultation with Seoul City and Jongno-gu officials resulted in the current plan to conserve and extend the hanok to the greatest extent feasible.

Preservation and Extension

As the plan was shifted toward preserving the hanok to the maximum degree possible, the main quarter with the main hall of the ⊏-shaped hanok was to be removed to transform it into an ∟-shaped hanok. A hybrid extension was carried out by integrating the remains of the hanok with a new structure in order to ensure the maximum floor area within the proposals of the time. Even though the two buildings were joined, they were separate structures, with only the outside walls connecting them. The yard from the old hanok was preserved, but it has been transformed into an area shared by the Oriental-style hanok and the Western-style building.

Reflection and Illusion in Half & Half Architecture

Although it's difficult to discern from the outside, the old hanok and the new architecture are joined together, making the interior space appear as if both buildings were sliced down the middle and then attached. Each half of the set is familiar, but they seem incredibly unusual and intriguing when standing side by side. In addition, the reversed perspective of the tiled roof reflected on the mirrored Barrisol ceiling of the pilotis area creates the illusion of a portal to another universe. The reflection of the hanok on the window panes facing the yard forms the illusion of a hanok lost in history,

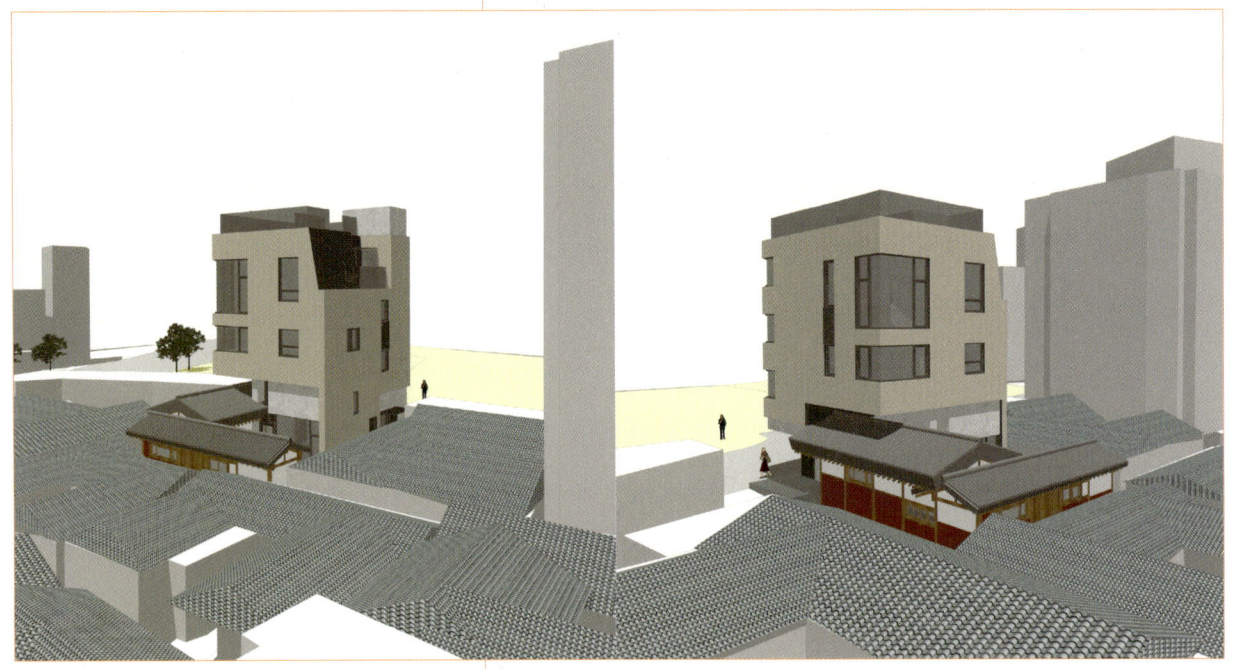

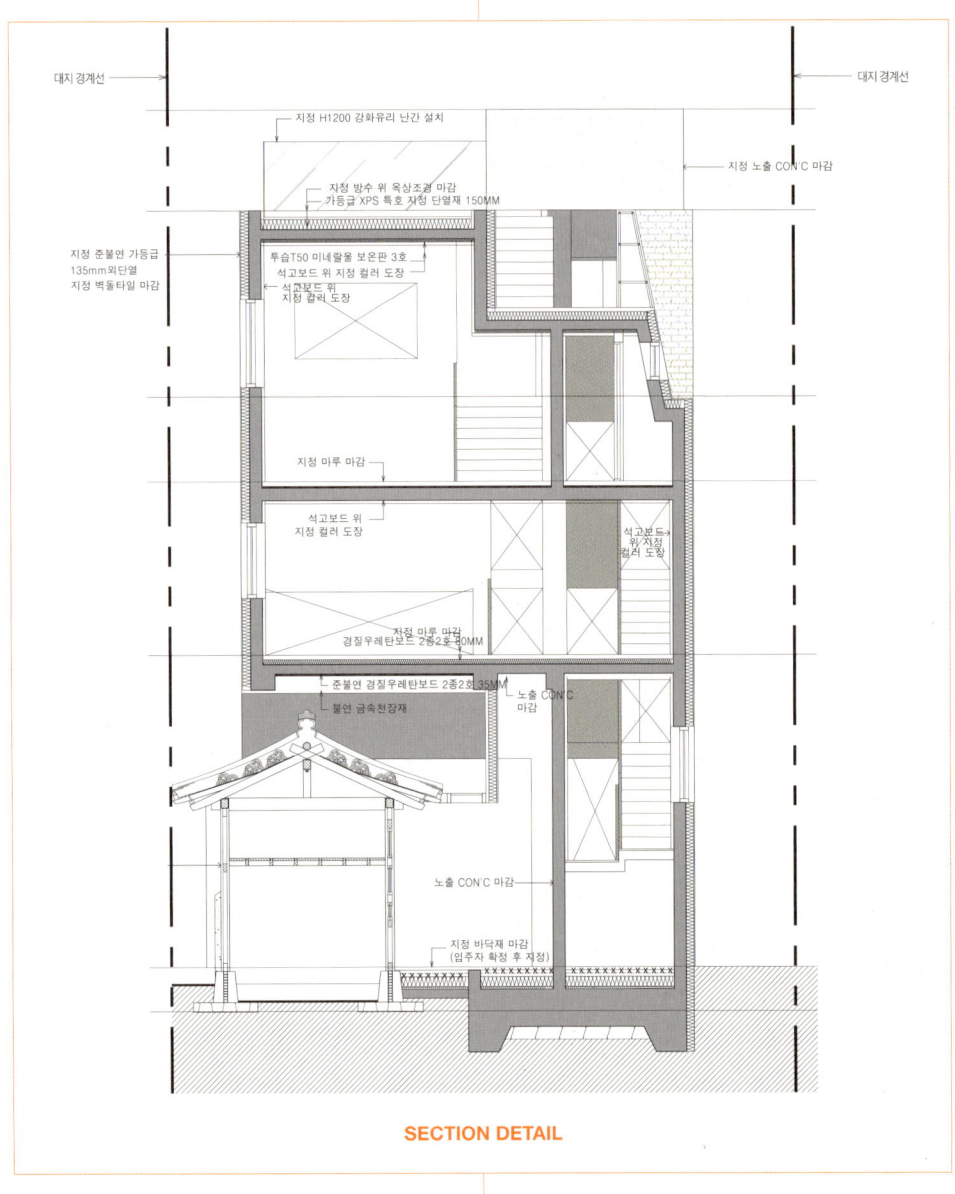

SECTION DETAIL

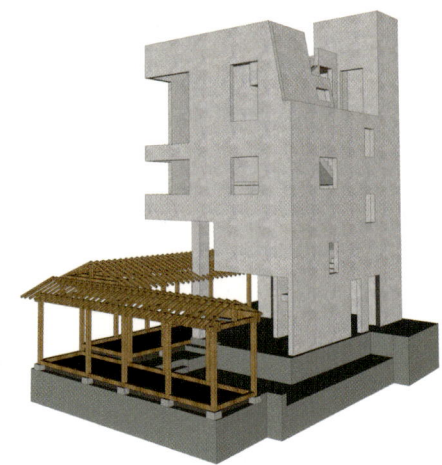

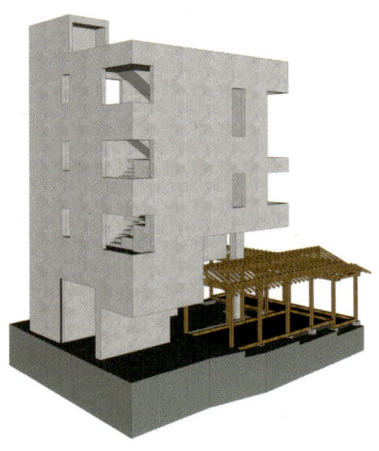

RENDERING

Neighborhood Facility

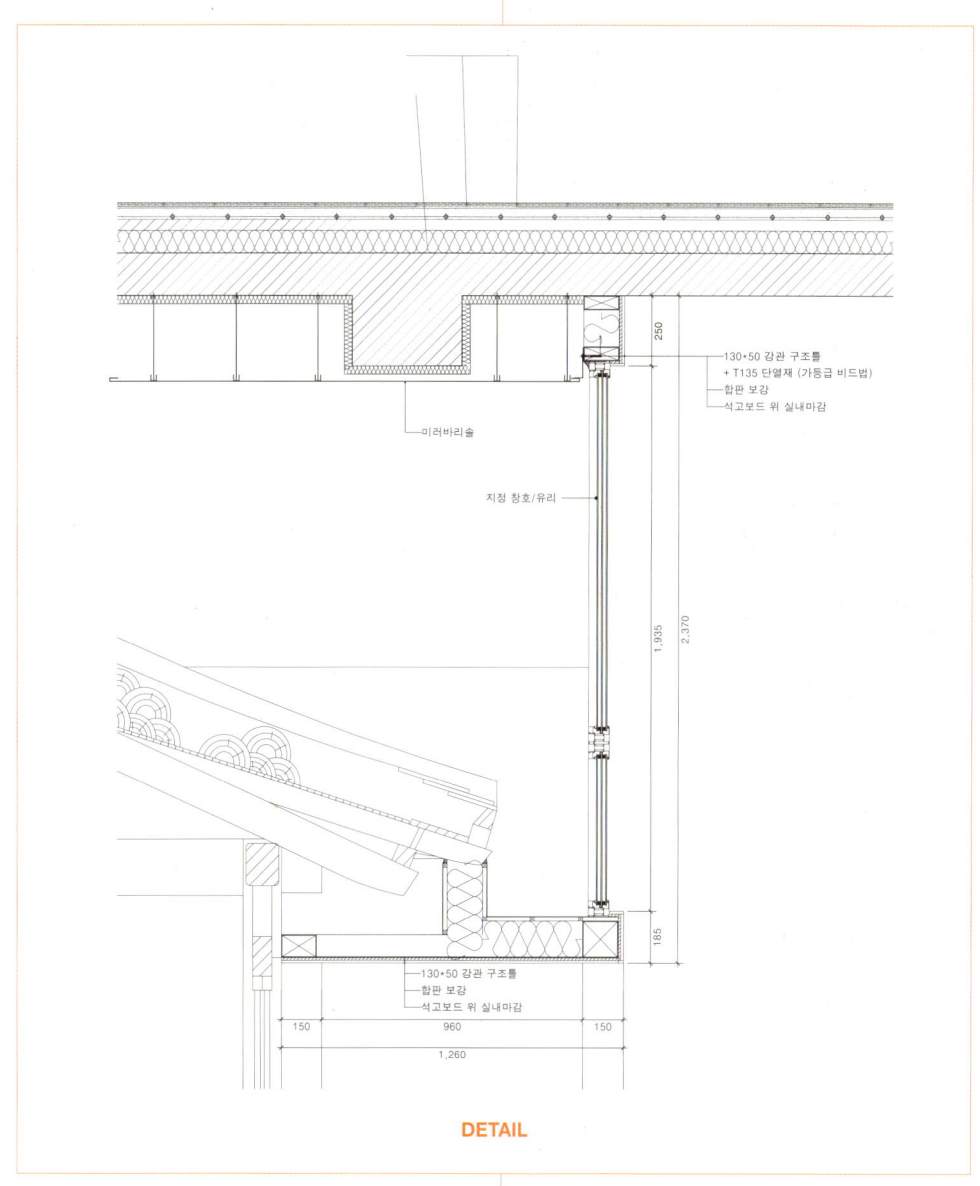

DETAIL

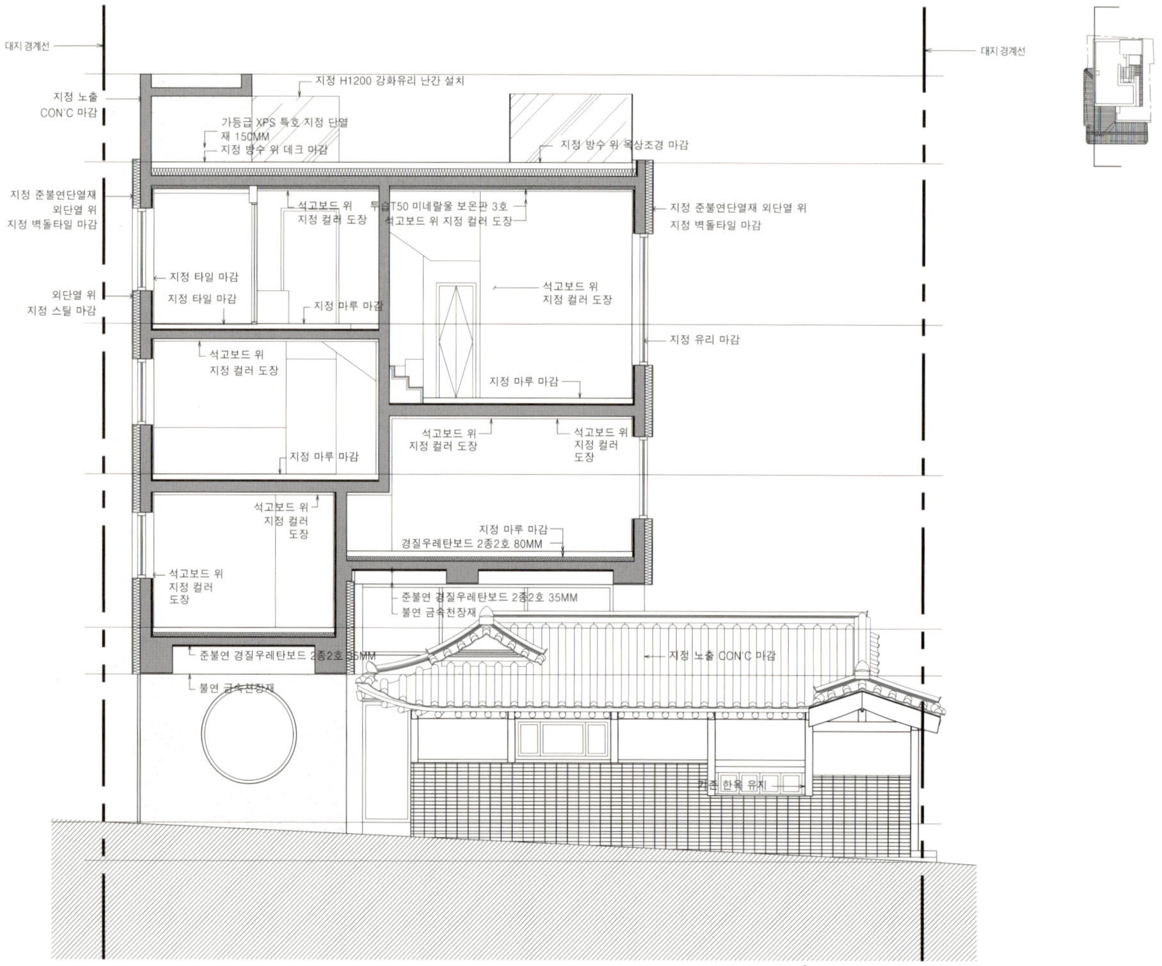

SECTION DETAIL

Neighborhood Facility

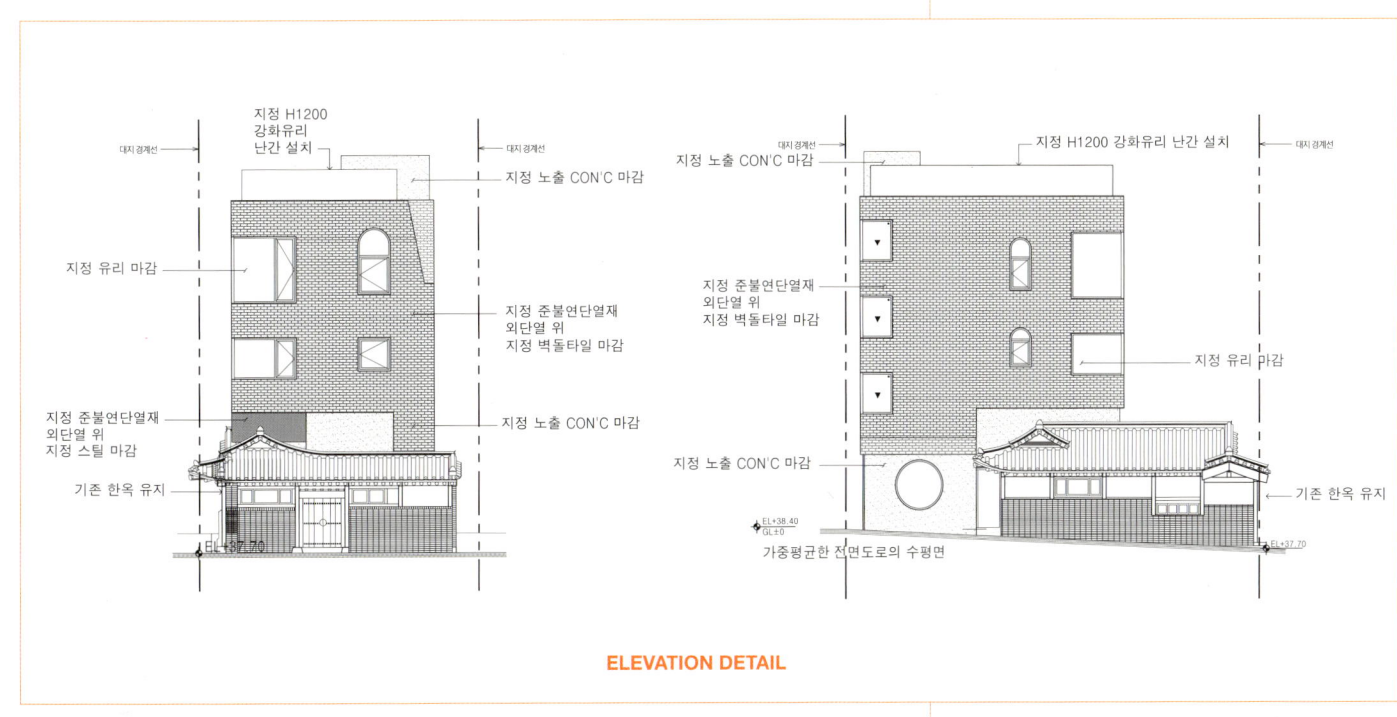

ELEVATION DETAIL

1 HANOK(TRADITIONAL KOREAN-STYLE HOUSE)
2 CAFE
3 ENTRANCE
4 LIVING ROOM
5 BALCONY
6 ROOM

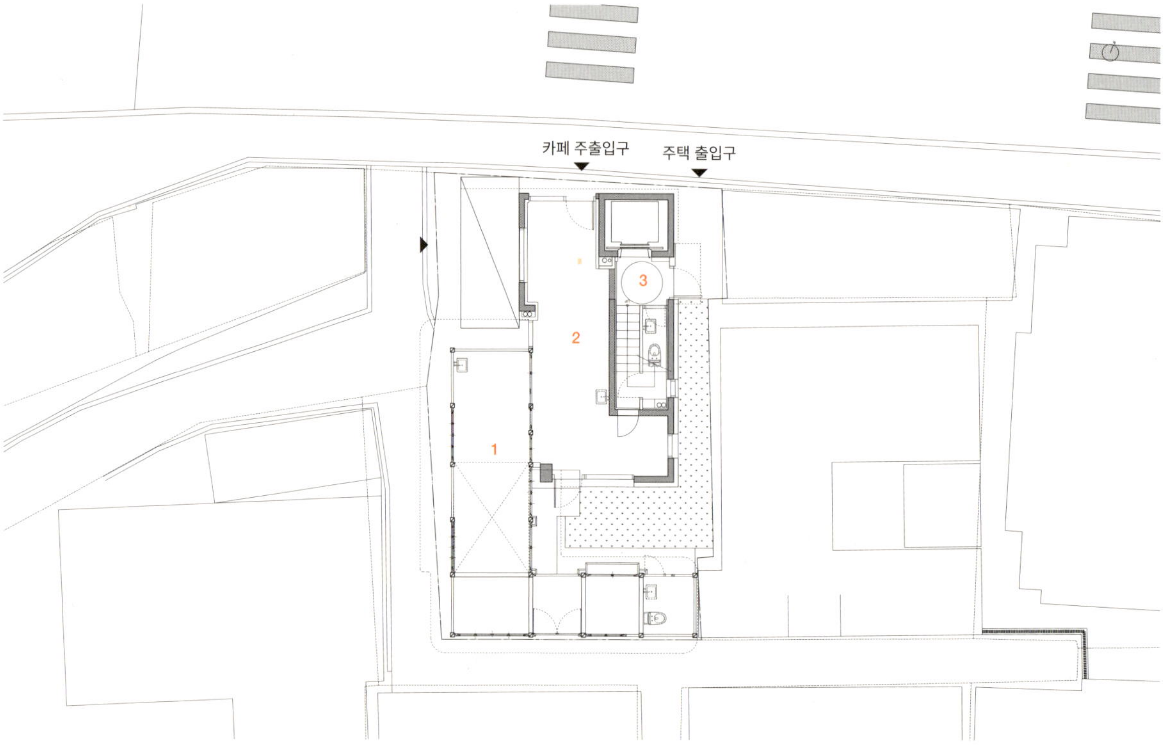

카페 주출입구   주택 출입구

1ST FLOOR PLAN

Neighborhood Facility

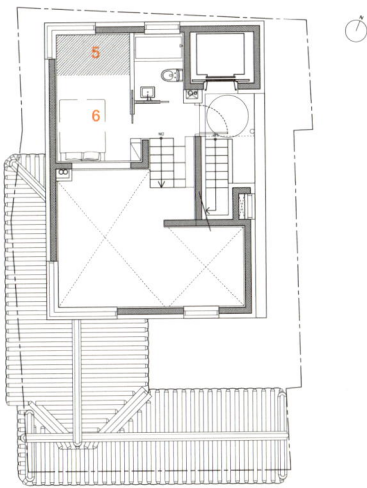

4TH FLOOR PLAN

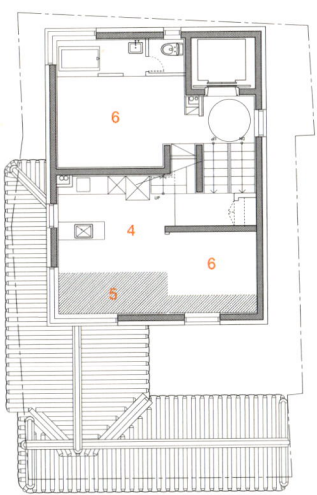

3RD FLOOR PLAN

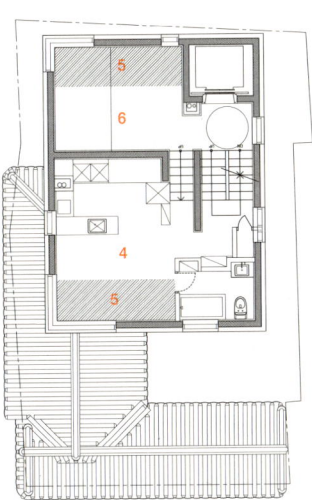

2ND FLOOR PLAN

giving us the impression that the past and the present have merged in front of our eyes.

A Western-style Structure on Top of a Hanok

Maintaining the hanok and satisfying the program and area requirements under a height restriction of 12 meters was a difficult task with an ordinary floor plan. A split-level floor plan, which would take care of the half-floor height difference due to the height of the hanok's roof, was naturally seen as an option, and a floor plan and space structure were proposed where every tiny space fits in like a jigsaw puzzle.

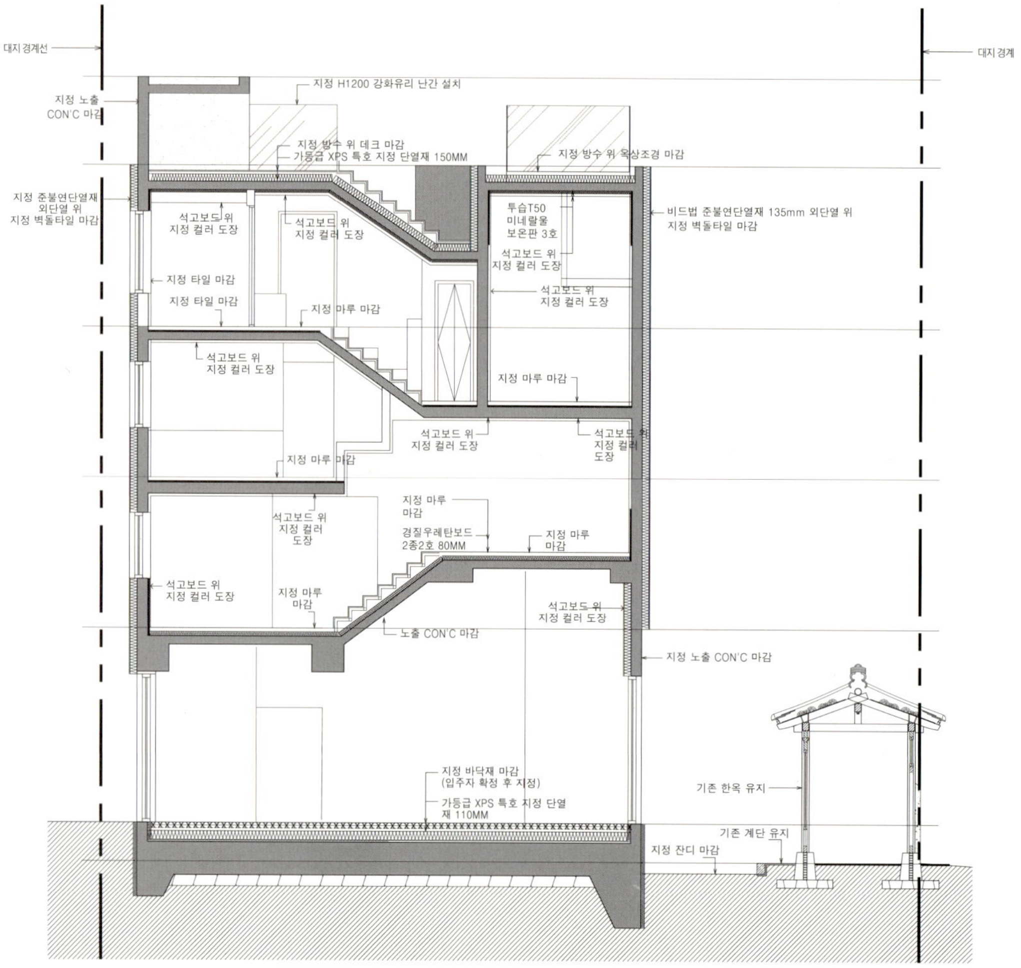

SECTION DETAIL

Neighborhood Facility

김윤수 / 바운더리스 건축사사무소
Kim Younsoo / Boundaries Architects

Boundaries Architects(바운더리스 건축사사무소)는 건축을 베이스로 하여 다양한 장르와의 교류를 통해 프로젝트를 진행하였으며, 경계를 넘나드는 오픈프로젝트를 기획하여 아티스트 및 디자이너의 자발적 참여를 통한 작업을 하고 있다. 2011년 '붉은 집', 'POP UP SPACE', 2012년 서울시 72시간 도시재생프로젝트 '서울채집', 2013년 세종문화회관 정오의 예술무대 'AUTUMN ART PROJECT', 'SOSO HOUSE' 2015년 '남원문화루' 등 다양한 작업을 협력하여 진행해왔다. 2016년부터 서울시 공공건축가/마을건축가로 활동하였으며, 2018년에는 '여주시 마을학교 마스터플랜'을 시작으로 '서해고 환경학교 프로젝트', '서울시 꿈담교실 건축가'로 활동하며 교육공간 건축의 변화에 대한 프로젝트들을 지속적으로 진행하고 있다. 또한 도서관, 어린이집, 커뮤니티 센터 등 다양한 공공 및 민간 프로젝트도 진행 중이다.

건축가 김윤수는 단국대학교 건축공학과를 졸업하고 경기대 건축전문대학원에서 건축 설계 전공 석사를 취득한 후 힘마 건축사사무소, 운생동 건축사사무소에서 다수의 프로젝트를 수행하였다. 2011년 바운더리스를 설립하여 건축, 인테리어, 설치 등의 작업을 진행하고 있다. 뿐만아니라, 공유주택, 공유업무공간인 WITHSOMETHING을 운영하며 이를 바탕으로 다양한 사회적 공유공간에 대한 실험을 하고 있다. 현재 용인시 수지구 동천동에 주차전용 건축물을 베이스로 다수의 건축물로 구성된 공동체주거 프로젝트의 디자인을 진행하며 새로운 1인 주거의 프로토타입 기획에 대한 자문을 하고 있다. 다양한 주거와 임대공간을 통한 변화의 방향과 지역 활성화를 통한 지역의 변화를 주시하고 있으며, 이를 실체화시키는 작업을 진행 중이다.

>> http://boundaries.co.kr

근린생활시설 7

101m² – 120m²

NEIGHBORHOOD FACILITY 7

AZARA 891 / 아자라 891
CENTRO CERO

KGA DESIGN STUDIO / KGA 디자인 스튜디오
Kreis Grennan Architecture

BIG STAIRS / 빅 스테어스
오후건축사사무소 + ㈜ 유타건축사사무소
OHOO ARCHITECTURE + UTAA COMPANY

CUP OF TEA ENSEMBLE / 차 한 잔의 앙상블
Kraft Architects

GOMBI / 곰비빌딩
㈜오엠엠건축사사무소 omm architects

SLEEPING LAB·ARCH / 슬리핑 랩-아치
Atelier d'More

OUTDOOR OFFICE / 아웃도어 오피스
ANDERS BERENSSON ARCHITECTS

Neighborhood Facility

Location
Buenos Aires, Argentina
Use
Multifamily Building
Site area
170m²
Built area
102m²
Total floor area
338m²
Floors
3F

Exterior finish
Exposed concrete, yellow microperforated sheet metal
Interior finish
Textured white walls
Project architect
Leonardo Gabriel Valtuille, Carolina Antolini, Pablo Javier Pugliese
Photographer
Javier Agustín Rojas

Azara 891

아자라 891

CENTRO CERO

Azara 891 is a low-density multifamily building located on a plot of land between party walls in Lomas de Zamora, south of Buenos Aires Metropolitan area. On an empty lot of limited dimensions (8.66m x 19.52m), it is prioritizing the preservation of a large existing avocado tree, adjacent to the southern neighbor.

Two functional units are defined per floor, which take the entire width of the lot, both in the front and in the back part of the building, separated by the Avocado Patio and the stairs on the north side. This arrangement allows multiple visual continuities, cross ventilation, and lighting.

The ground floor is free to the front for pedestrian access and garages and is completed with a flat with its own backyard. The stairs and the hallways are related to the patio. The third floor allows each unit to have its individual terrace with a barbecue to reformulate its own patio in height.

The exposed concrete, the large edge-to-edge windows, the textured white walls, and the linear space generated by the tensioners that will receive vegetation, give the whole a contemporary and stripped-down aesthetic. The yellow microperforated sheet metal gate illuminates and stands out by contradiction, provides privacy as well as visual permeability depending on the time of day. The material continuity of the pavement of the sidewalk dissolves the limits reinforcing the idea of prolonging the public over the private, improving the pedestrian experience of the neighborhood.

Neighborhood Facility

SITE PLAN

FÉLIX DE AZARA

LAS HERAS

175

Neighborhood Facility

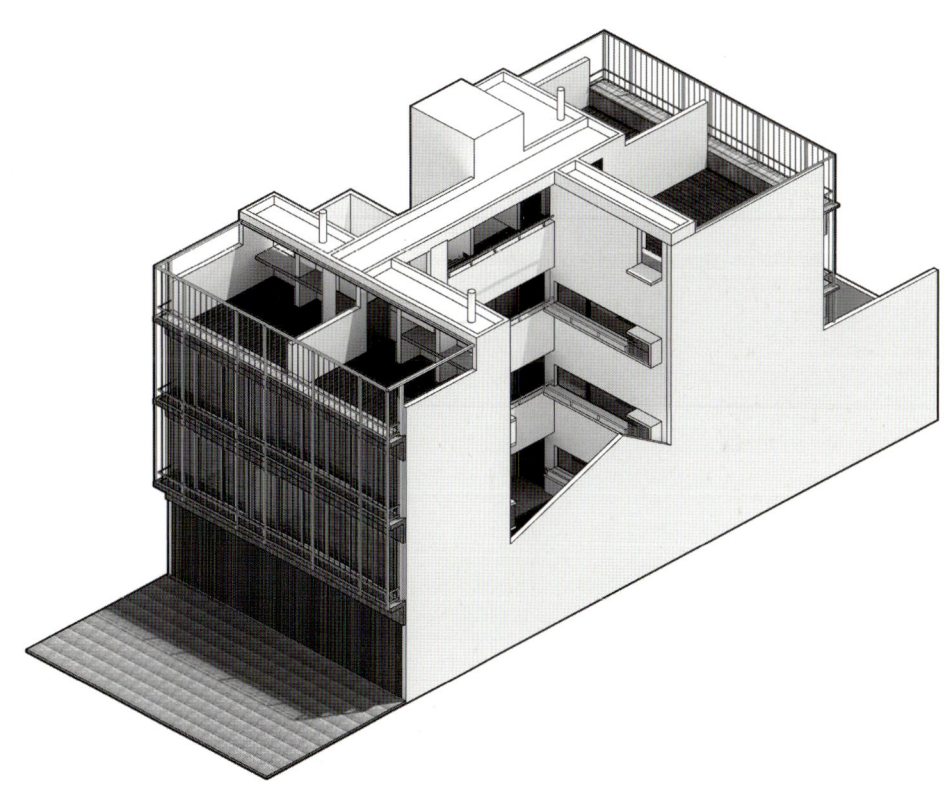

AXONOMETRIC

아자라 891(Azara 891)은 대도시 부에노스아이레스 남쪽에 위치한 로마스데사모라(Lomas de Zamora)에 있는 파티 벽 사이 부지에 지은 저밀도 다가구 빌딩이다. 제한된 크기(8.66m x 19.52m)의 공터에는 남쪽 이웃집과 가까이에 있는 기존의 큰 아보카도 나무가 우선적으로 보존되어 있다.

층당 2개의 기능 단위가 정의되며, 이는 빌딩의 전면과 후면 모두에서 부지의 전체 너비를 차지하고, 아보카도 파티오와 북쪽 계단에 의해 분리된다. 이러한 배치를 통해 다중 시각적 연속성, 교차 환기 및 조명이 가능하다.

지상층은 보행자의 출입과 차고 용도를 위해 전면을 향해 자유롭게 열려 있으며, 자체 뒷마당을 갖추었다. 계단과 복도는 파티오와 연결되어 있다. 3층의 각 유닛에는 바비큐 장비를 갖춘 개별 테라스가 있어 자체 파티오 높이를 재구성할 수 있다.

노출된 콘크리트, 끝에서 끝으로 이어지는 큰 창문, 질감 있는 하얀 벽, 식물을 받을 텐셔너에 의해 생성된 선형 공간은 전체적으로 현대적이고 절제된 미학을 선사한다. 노란색 미세 천공 판금 출입문은 빛이 나고 대비로 인해 눈에 잘 띄며 시간대에 따라 시각적 투과성은 물론 프라이버시까지 제공한다. 보도 포장의 물질적 연속성은 경계를 해소하여 개인보다는 공공성을 연장한다는 아이디어를 강화함으로써 동네의 보행자 경험을 향상시킨다.

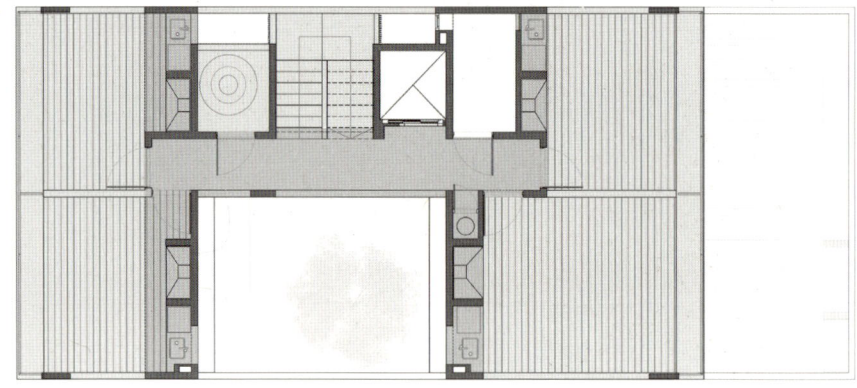

3RD FLOOR PLAN

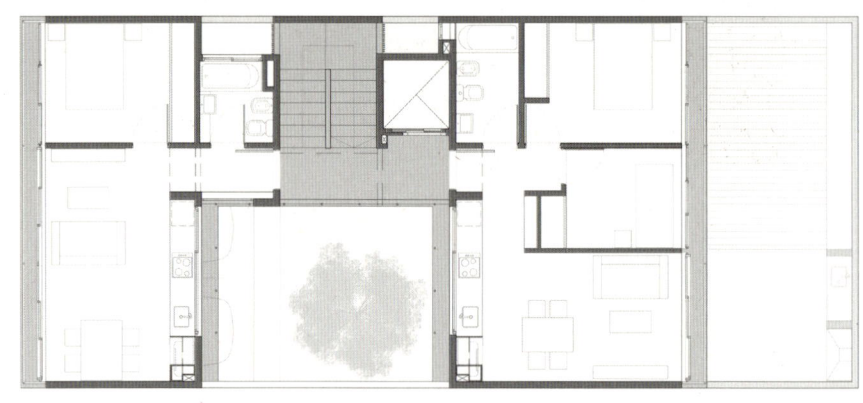

TYPICAL FLOOR PLAN

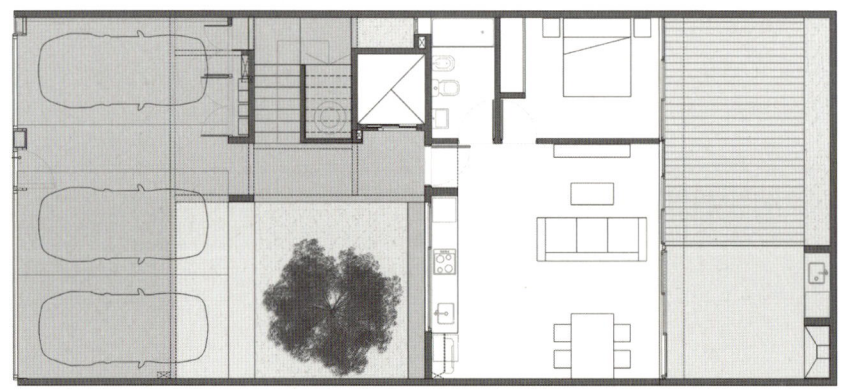

GROUND FLOOR PLAN

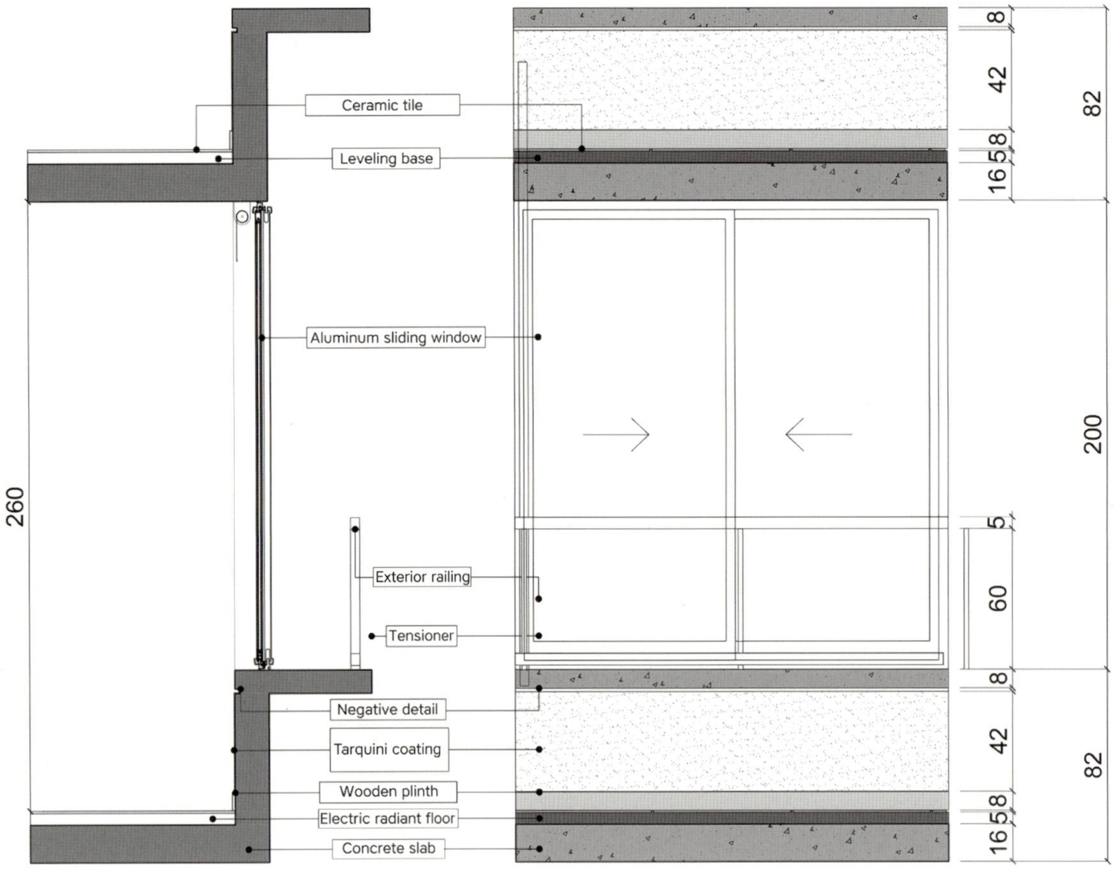

CONSTRUCTIVE DETAIL

Neighborhood Facility

SECTION

ESTUDIO CENTRO CERO

ESTUDIO CENTRO CERO was born as a flexible and interdisciplinary platform that promotes the linking of professionals in specialized work groups with precise objectives. In this way they manage to cover different scales of integral management of architectural works, maintaining high levels of dedication to each assignment.

Their experience is based on a revision of the traditional architectural firm and its forms of commercialization. They aim to look for totalizing ideas in the management of the work and include the study of the land, the environment, the project and management and the construction.

They base our practice on research, teaching (Universidad de Buenos Aires) and constant training. They take each production chance as a milestone of thought on contemporary concepts and techniques.

>> www.centrocero.com.ar

Neighborhood Facility

Location
Sydney, Australia
Use
Office, Residential
Site area
160m²
Built area
102m²
Total floor area
161m²
Floors
2F

Exterior finish
Face Brick, Timber Cladding (Supplier: ABODO)
Interior finish
Rendered Walls, Plasterboard, Timber Floors
Project architect
Christian Grennan, Lin Lee
Photographer
DOUGLAS FROST & ANDREAS BOMMERT

KGA DESIGN STUDIO

KGA 디자인 스튜디오

KREIS GRENNAN ARCHITECTURE

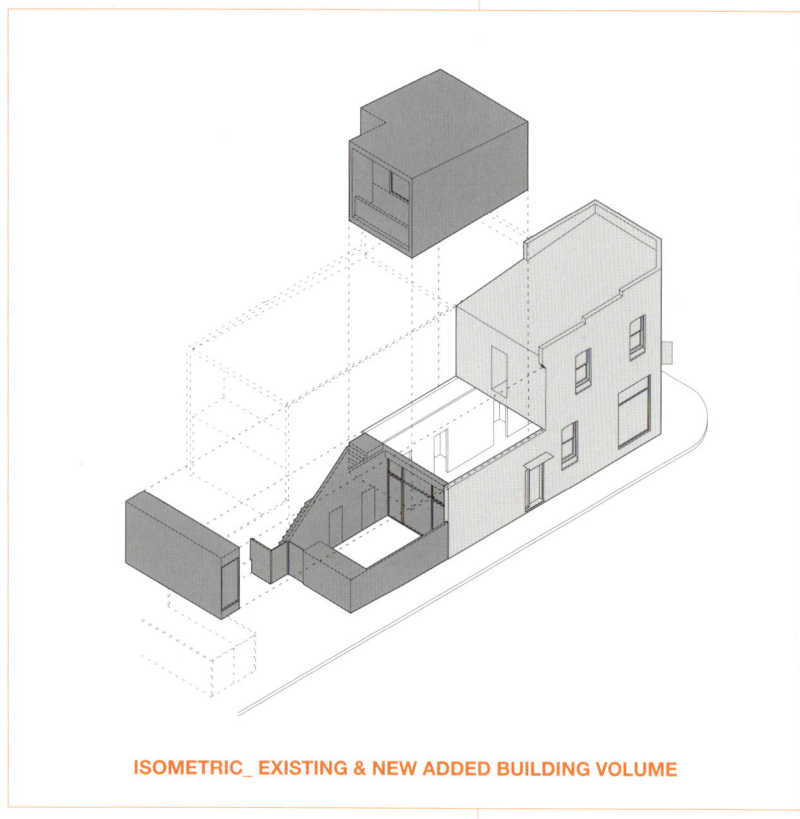

ISOMETRIC_ EXISTING & NEW ADDED BUILDING VOLUME

Neighborhood Facility

An adaptive re-use of a period corner shop into a shop-top, accommodating a design studio on the ground floor and a residential apartment on the first floor, with private open space and parking facilities at the rear. The design intent was to add a contemporary volume that highlighted and respected the existing building creating two separately strong building elements forming one new strong composition. The stripped brick walls, showing traces or previous uses, contrast with the restrained black board & batten timber cladding.

The dramatic cantilevered balcony provides a sun filled open space for the apartment, and shading for the commercial courtyard below. The external hard surfaces and car parking are paved in permeable brick pavers, allowing rainwater to naturally absorb into the ground, irrigating the garden and trees and reducing water runoff.

시대상이 묻어나는 길모퉁이 가게를 주거 겸용 오피스로 조정 및 재사용하였다. 1층에는 디자인 스튜디오를, 2층에는 주거용 아파트를 조성하고 뒤쪽에는 프라이빗한 공터와 주차공간을 마련했다.

설계 의도는 현대적인 볼륨을 추가하여 하나의 새로운 강력한 구성을 형성하는 뚜렷한 두 개의 개별적 빌딩 요소를 생성하는 기존 건물을 강조하고 그에 대한 존중을 나타내고자 함이다. 이전 사용의 흔적이 묻어나는 벗겨진 벽돌 벽은 절제된 색상의 검은 판자와 널빤지로 된 목재 피복 소재와 대조를 이룬다.

드라마틱한 캔틸레버 발코니는 아파트의 햇살 가득한 열린 공간이며 그 아래 사무실 정원에 그늘을 드리운다.

외부의 딱딱한 표면과 주차장은 침투성 벽돌 재료로 포장되어 빗물이 자연스럽게 땅속으로 흡수될 수 있도록 하여 정원과 나무를 관개하고 물의 유출을 줄인다.

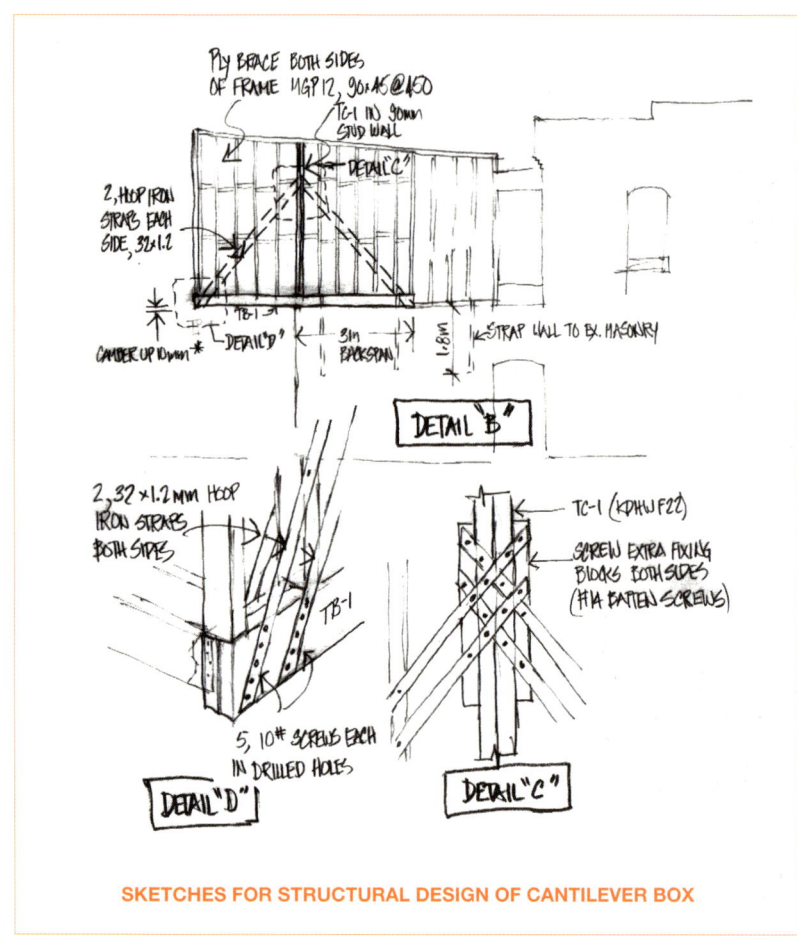

SKETCHES FOR STRUCTURAL DESIGN OF CANTILEVER BOX

TIMBER CLADDING

Neighborhood Facility

1 ENTRY
2 MEETING ROOM
3 STUDIO
4 PRINTER ROOM
5 KITCHENETTE
6 TOILET
7 STORAGE
8 BATHROOM
9 APARTMENT ENTRY
10 COURTYARD
11 BIN STORAGE
12 PARKING
13 BALCONY
14 DINING
15 LIVING
16 KITCHEN
17 BATHROOM
18 STUDY
19 BEDROOM

1ST FLOOR PLAN

GROUND FLOOR PLAN

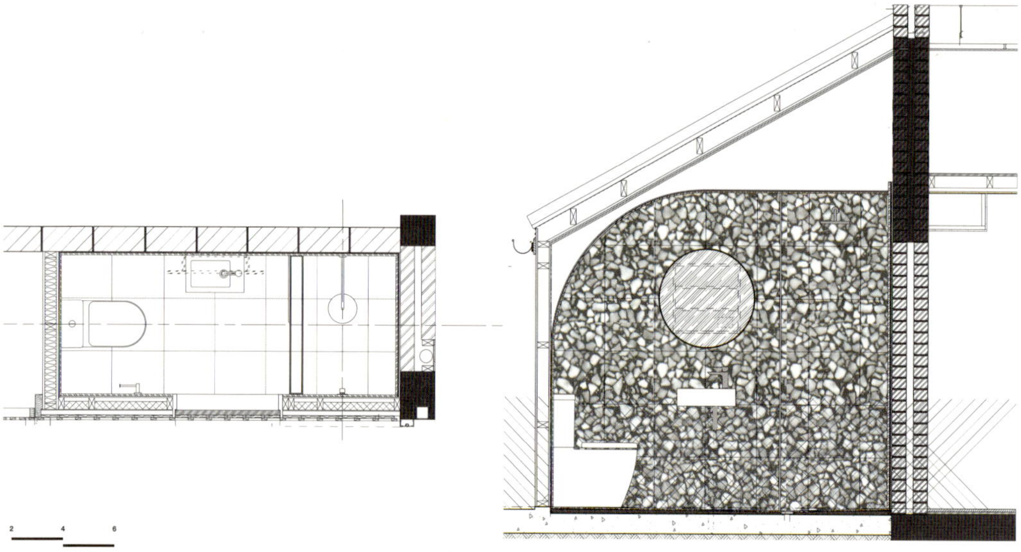

BATHROOM DETAIL

1 PARKING
2 GARDEN
3 STUDIO
4 MEETING ROOM
5 BALCONY
6 KITCHEN
7 BATHROOM
8 STORAGE
9 BEDROOM

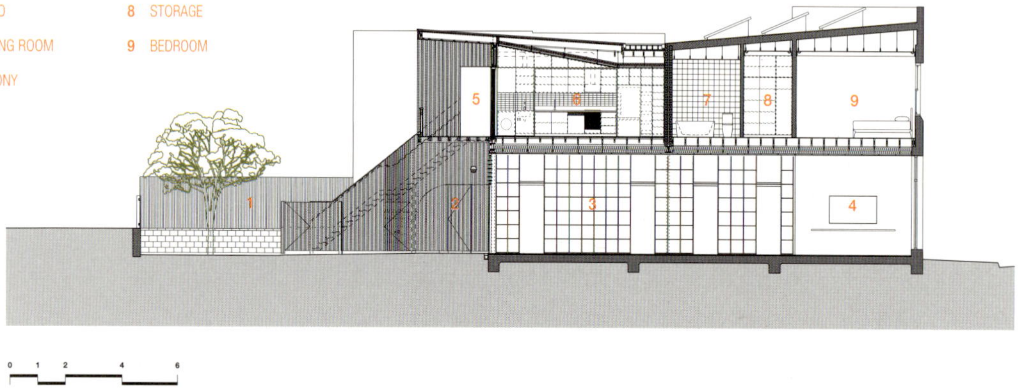

SECTION

Neighborhood Facility

KREIS GRENNAN ARCHITECTURE

WE LISTEN, WE CREATE, WE CRAFT GREAT SPACES

KGA has a core team of senior staff with over 25 years of experience in architecture
and interior design. We are specialised in residential housing design but are experienced in most other building typologies.

The studio's design process is always collaborative - listening carefully to clients, consultants and crafts people; asking questions, thinking critically and providing a transparent decision-making process. Our commitment to design quality is matched by our attention to detail. The design process is driven and tested by six design ethos principles:

1. Resolve complex design issues through coherent FORM
2. Enhance space & amenity by celebrating natural LIGHT
3. Work with the inherent beauty of natural MATERIALS
4. Integrate & evaluate SUSTAINABILITY
5. Augment location through considered PLACEMENT
6. Complement usability & delight by understanding USERS

\>> www.kreisgrennan.com.au

Location
Gangnam-gu, Seoul, Republic of Korea
Use
Store
Site area
182.10m²
Built area
107.78m²
Total Floor area
559.39m²
Floors
B2, 5F

Exterior finish
Mono White Cream Tile, STO
Interior finish
Exposed concrete
Project architect
Kim Chang Gyun, Bae Young Sik, Kim Haahlyn
Construction
STARSIS
Photographer
Bae Jihoon

BIG STAIRS

빅 스테어스

오후건축사사무소 + ㈜유타건축사사무소
OHOO ARCHITECTURE + UTAA COMPANY

논현동 주택가는 밀도 높은 주거지로서 오랜 시간 자리를 지켜온 주택들이 하나 둘씩 철거되고 각양각색의 새로운 건물이 들어서고 있다. 점차 상권이 확장되고 빠르게 변화하는 시기에 맞춰 거리가 변하고 있는 것이다. 그 틈 속에 논현동 'BIG STAIRS'가 자리잡고 있다.

부지 일대는 가파른 경사가 있고 차량 통행이 많아 어지러이 뻗어있는 전신주처럼 유난히 혼잡하다. 복잡한 공간일수록 단순한 매스와 단색의 입면 재료가 오히려 눈길을 끌게 한다. 논현동 'BIG STAIRS'는 최대한 색감을 배제하고 무채색계열의 백색타일을 사용하여 어수선한 거리에 정돈된 차분한 인상을 주고자 했다. 또한 백색타일의 패턴을 층마다 다르게 담아내어 지나가는 사람들에게 색다른 인상을 주고 호기심을 자극할 수 있는 건물이 되고자 한다.

Neighborhood Facility

접근성과 가시성에 따라 상가의 임대 비용이 달라지기 때문에 떼려야 뗄 수 없는 상관관계이다. 상가는 지면과 가까울수록 다른 층에 비해 접근성이 좋기 때문에 임대료가 높게 측정된다. 그렇기 때문에 사람들에게 어떻게 하면 건물로 자연스럽게 접근하도록 유도할지 생각하는 것은 건축가의 오랜 고민이다. 논현동 부지는 경사지인 대지의 조건을 활용하여 지하층을 도로에서 쉽게 인지할 수 있게 노출시켜 1층과 다를 바 없는 쾌적한 공간이 되도록 했다. 2층 또한 상부층으로 가는 외부계단이 도로에서 시각적으로 존재감을 드러내 자연스럽게 발길이 연결되어 이어지도록 했다. 저층부는 시원하게 비워낸 공간은 도시의 가로풍경을 느슨하게 열어줘 거리에 여유로움을 제공한다.

주거지역의 땅은 일조권 사선제한 법규로 인해 상부층으로 갈수록 건물이 줄어들게 된다. 이 부지의 경우에도 최상층이 일조권 사선제한 영향으로 실공간이 10평 남짓 되지 않아, 일반적인 엘리베이터와 계단실이 묶여있는 코어로 계획하기에 부적합했다. 이러한 문제를 해결하기 위해 일조권 사선제한으로 비워진 공간에 외부계단을 계획하고, 계단하부를 내부 창고로 활용하여 조금이나마 전용공간을 확보했다. 이 계단은 오르내리며 도시풍경을 즐길 수 있고 트인 시야를 바라보며 여유를 즐길 수 있는 공유 테라스가 되기도 한다. 공유 테라스는 바쁜 일상 속에 잠시나마 바람을 맞고 계절을 느낄 가치 있는 공간이길 기대한다.

Neighborhood Facility

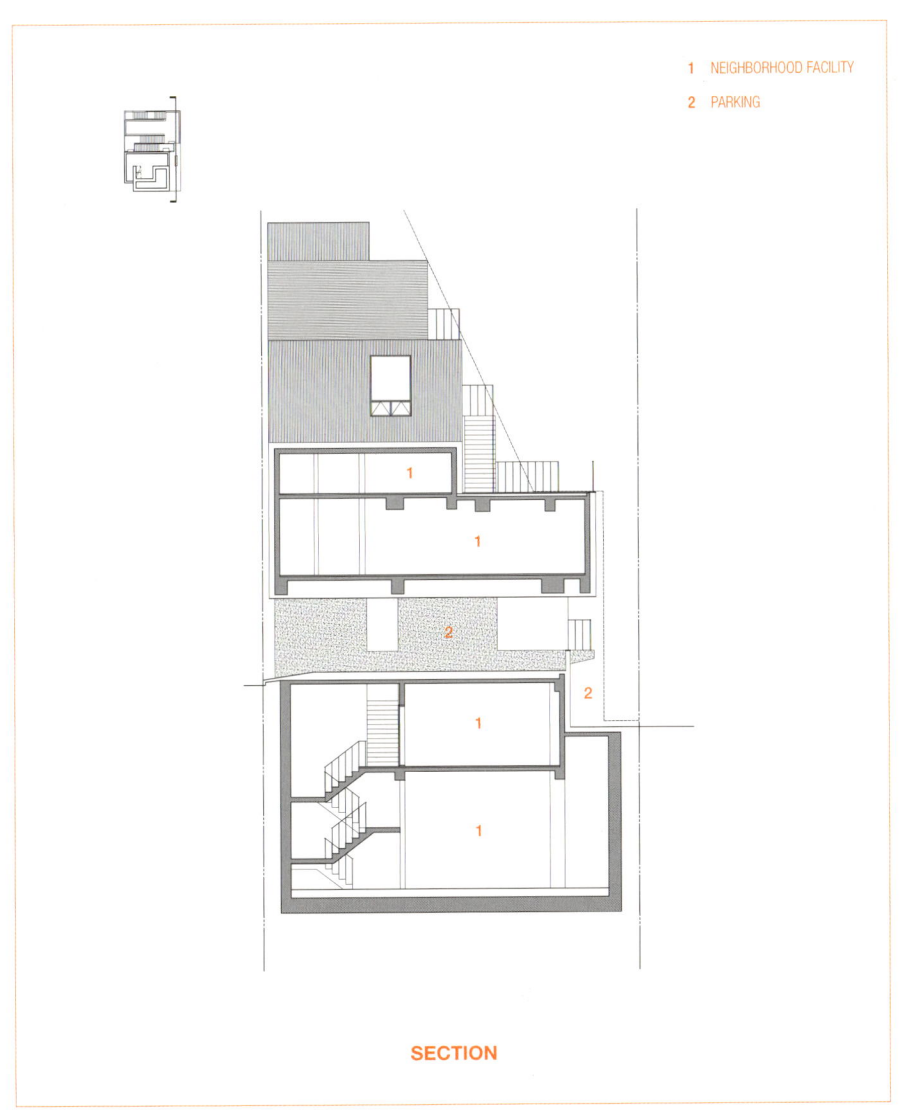

| 1 | NEIGHBORHOOD FACILITY |
| 2 | PARKING |

SECTION

The residential area in Nonhyeon-dong is densely populated, where houses that have been there for a long time are being demolished one by one and new buildings of all shapes and sizes are being erected. Streets are evolving in tandem with the rapid pace of change and the steady expansion of commercial districts. Nonhyeon-dong's "Big Stairs" is standing in the middle of it all.

Because of a steep slope and heavy traffic around the site, it appears extremely congested, as if power lines stretch out from the utility poles in all directions. In a complex area, the simple mass and single-color elevation material can be more eye-catching. The "Big Stairs" in Nonhyeon-dong attempted to create an impression of order and tranquility in the cluttered streets by avoiding bright colors and employing off-white-colored tiles. In addition, we aimed to build a structure that would give a unique appearance to pedestrians and attract their interest by capturing varied patterns of white tiles on each floor.

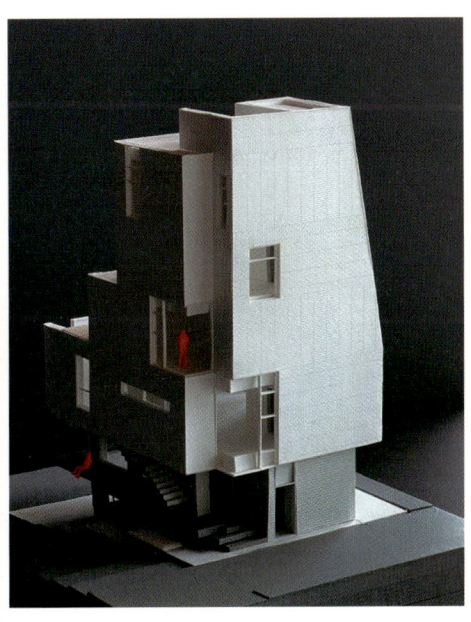

Accessibility and visibility are inextricably correlated because the rental prices of commercial buildings are influenced by them. The closer a store is to the ground, the higher its rent because the accessibility is better than on other floors. Therefore, an architect's primary concern has always been how to encourage people to approach the building. The Nonhyeon-dong site utilizes its condition of being on a slope to make the basement level visible from the streets and make it as attractive as the first floor. The exterior stairs leading to the second and other upper floors visually reveal their presence on the road, seamlessly connecting visitors' movements to the upper levels. The vacant spaces on the lower levels become part of the city's streetscape, giving the location a sense of openness.

Due to the laws regulating the slant line for daylight, buildings in residential areas tend to have narrower upper levels and the slant line restriction at this site means the actual area on the top floor was less than 33m^2, which made it unsuitable for planning as a core with a set of an elevator and a stairwell. To address this issue, an exterior staircase was built in the space abandoned by the restriction of the slant line for daylight, and the bottom part of the stairs was converted into an interior storage area to increase the amount of functional space. These stairs become a communal terrace where people can enjoy the cityscape and relax in the open view while moving up and down. The shared terrace is anticipated to be a valuable space where people can feel the wind and the season, even for a short moment in their hectic daily lives.

1 NEIGHBORHOOD FACILITY
2 TOILET
3 MACHINE ROOM
4 PARKING

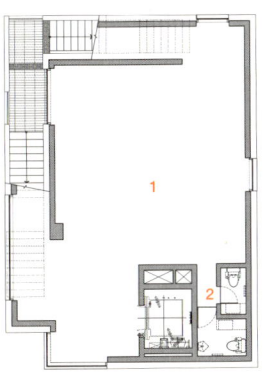
2ND FLOOR PLAN

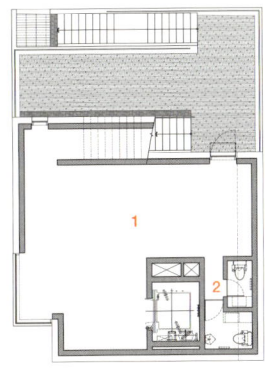
3RD FLOOR PLAN

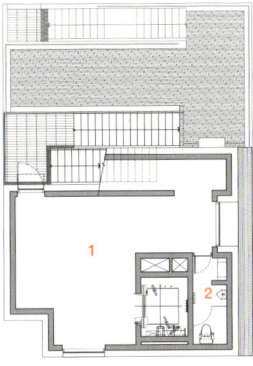
4TH FLOOR PLAN

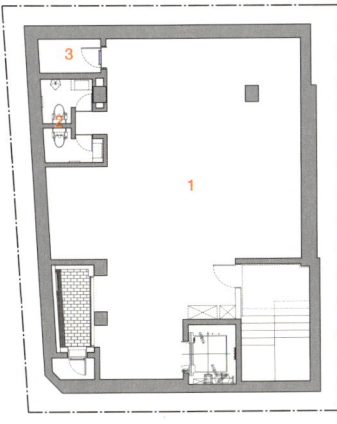
B2 FLOOR PLAN

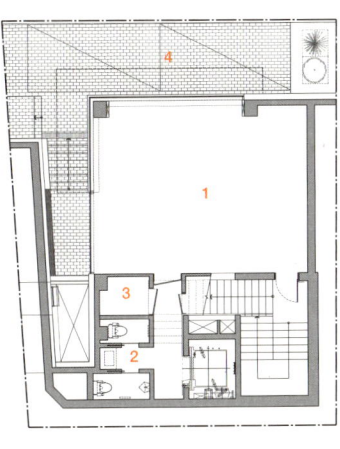
B1 FLOOR PLAN

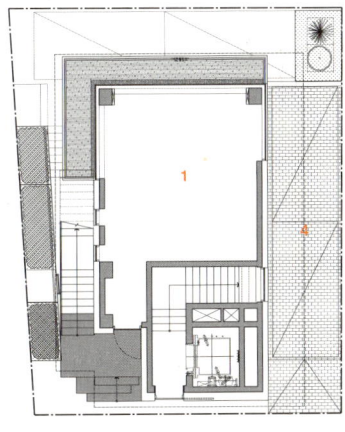
1ST FLOOR PLAN

1. NEIGHBORHOOD FACILITY
2. TOILET
3. STORAGE

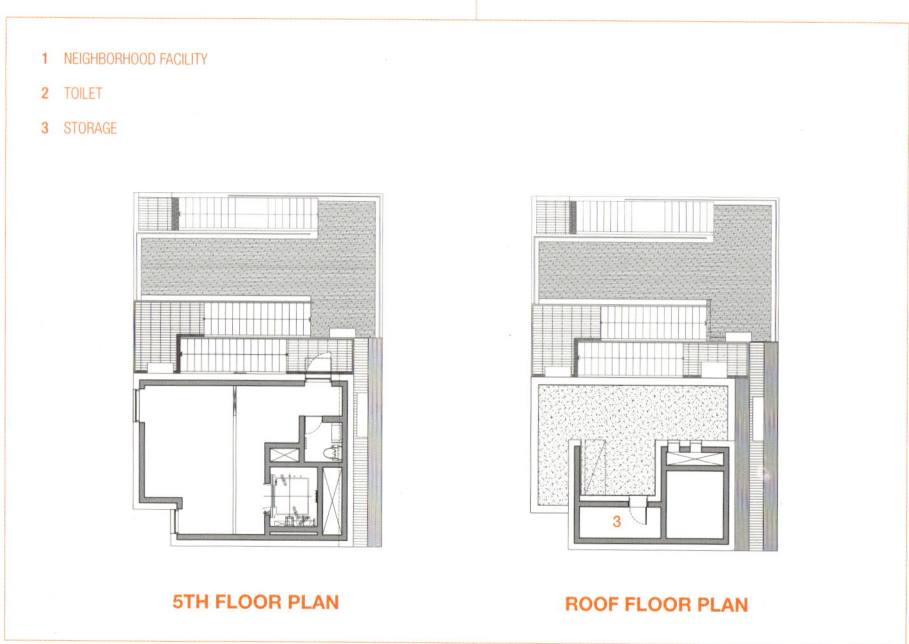

5TH FLOOR PLAN   **ROOF FLOOR PLAN**

Neighborhood Facility

1	NEIGHBORHOOD FACILITY
2	MACHINE ROOM
3	TOILET
4	TERRACE

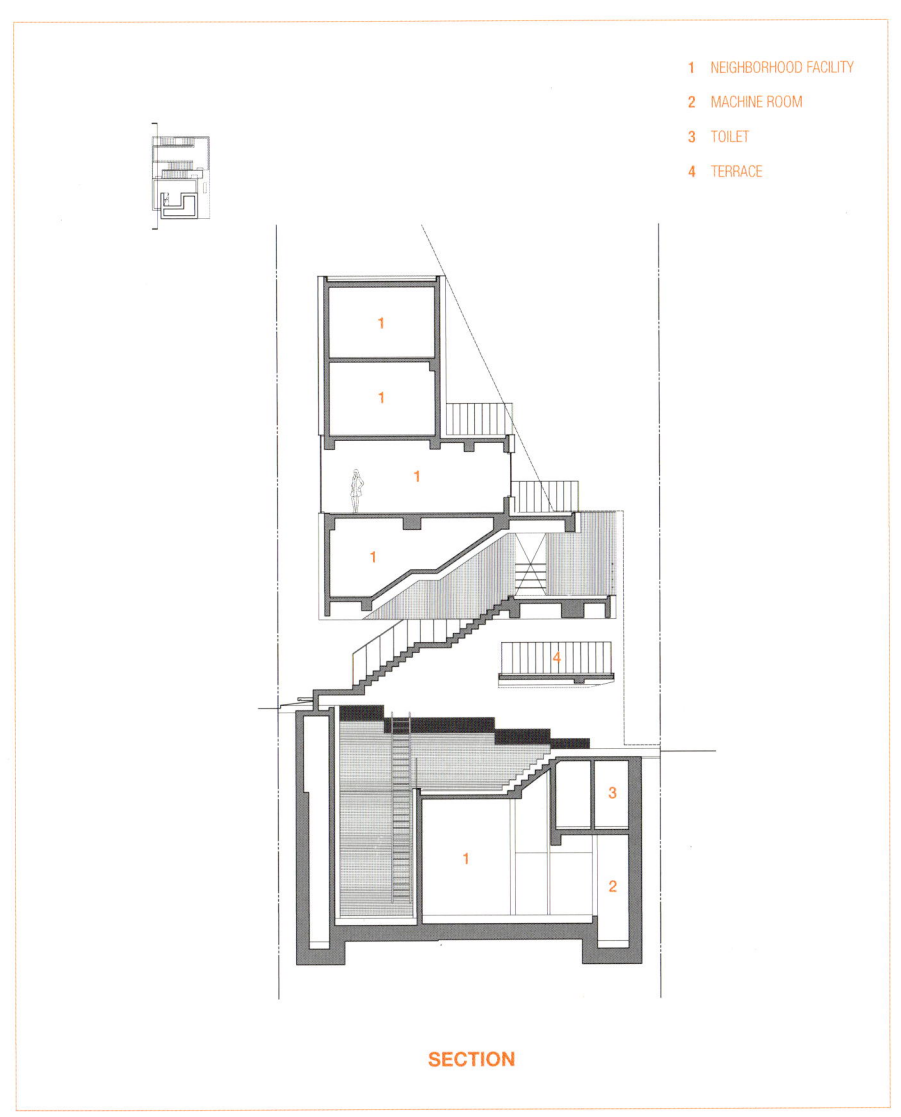

SECTION

오후건축사사무소 + ㈜ 유타건축사사무소
OHOO ARCHITECTURE + UTAA COMPANY

노서영, 김하아린 / 오후건축사사무소

건축가 노서영과 김하아린은 서울시립대학교 건축학과, 동아대학교 건축학과를 졸업하고 유타건축사사무소에 입사해 다양한 작업에 참여하여 실무 경험을 쌓은 뒤 오후건축사사무소를 개소했다. 대표 작업으로 상암동 'RED HOLE' 근린생활시설, 식금리 '休園' 연수원, 등촌동 '오각' 상가주택, 제천 단독주택 등이 있다. 사람과 공간, 그것과 관계를 맺는 일상 모든 것들이 한데 어우러지는 모습을 상상하며 그 속에서 건축적 고민과 시도를 지속하고자 한다. 어려운 건축 담론을 떠나 일상 속 아름다운 공간을 구현하고 도시조직에 긍정적인 변화를 유도하고자 한다.

>> www.ohooarch.com

김창균 / 유타건축사사무소

1971년생으로 서울시립대학교 건축공학과를 졸업하고 동대학원에서 석사 학위를 받았다. 해병대사령부 건축설계실, 에이텍건축 등에서 손 도면으로 시작하여 건축설계뿐 아니라 다양한 작업에 참여하며 실무경험을 쌓았고, 2006년 (주)리슈건축사사무소 공동대표를 거쳐 2009년 UTAA건축사사무소를 개소하였다(한국건축사). 현재 (주)유타건축사사무소 대표로, 서울시 공공건축가이며, 문화체육관광부에서 주관하는 '젊은 건축가 상'을 2011년 수상한 바 있다. 주요 작업으로 포천 피노키오 예술체험공간, 서울시립대학교 정문, 삼청가압장, 수원 상가주택 (The Square), 울산 간절곶 카페0732, 운중동 단독주택(도시채), 세종 단독주택(하품집) 등이 있다.

>> utaa.co.kr

Location
Hida Takayama, Gifu Prefecture, Japan
Use
Hotel+multipurpose lounge
Site area
131.70m²
Built area
112.28m²
Total floor area
324.90m²

Structure
RC
Floors
3F
Exterior finish
Plasterer, galvalume steel sheet spandrels, Thinned cedar wood + (Japanese) persimmon tannin
Interior finish
Thinned cedar wood + (Japanese) persimmon tannin, Sancyu washi paper, thinned cedar flooring, carpet, Hirari Sato

Project architect
Atsushi Nakamura, Hirari Sato
Photographer
Masao Nishikawa / Munetaka Onodera

Cup of Tea Ensemble

차 한 잔의 앙상블

Kraft Architects

문화소비형 관광에서 문화생산형 관광으로

본 프로젝트는 히다 다카야마에 위치한 작은 호텔이다. 이 지역은 연간 약 400만 명의 관광객이 찾는 인기 관광지로, 시라카와고의 세계유산과 오래된 거리가 유명하다. 이번 프로젝트에서는 역사적 사실을 추적하는 전통적인 관광과 함께 미래를 향해 나아가는 문화 창출 허브로서의 호텔이라는 아이디어를 추구함으로써 히다 다카야마를 향후 100년 문화로 이어지는 살아있는 지역적 가치의 출발점으로 만들었다.

현대화 이후 지속 가능한 방식의 지역주의를 어떻게 창출할 것인가?

모더니즘 건축이 시작된 지 약 100년이 지났다. 오늘날 현대화의 혜택으로 우리는 도시나 시골 그 어디에 있든 가능한 한 동일한 서비스와 생활 방식을 즐길 수 있다. 그러나 산업화된 자재로 지은 빌딩과 공간은 지역의 고유성을 잃었다. 앞으로의 100년을 내다보며 현대화를 수용한 도시의 고유성을 지속 가능한 방식으로 도출하는 방식을 통해 현재 우리의 생활, 환경과 상호작용하는 방식 자체가 미래 유산으로서의 가치를 가지고 전승될 수 있도록 하고자 한다.

지역의 고유성을 창출하기 위한 "초소형-주택 양식"

모더니즘으로 인해 획일화된 지역의 고유성을 다음 세대를 위해 되살리기 위해 건축이 할 수 있는 일이 무엇일지 생각하면서, 나는 이 지역을 구성하는 가장 작은 단위를 가까이에 이미 존재하는 것들로 재구성하는 작업부터 시작했다. 지역 고유의 주택 양식을 최단 거리에서 실현할 수 있는 규모로 세분화하는 것을 "초소형-주택 양식"이라고 불렀고, 경제와 법률을 포함한 사회적 시스템의 더 큰 틀에 성공적으로 통합될 수 있고, 환경과 사회적 순환 내에서 건설될 수 있는 새로운 형태의 건축물을 만드는 것을 목표로 했다. 건축적 행위란 "있는 것을 최대한 활용하는 것"이며, 이를 풍요로운 숲과 함께 하는 생활문화와 연결하여, 최대한 짧은 거리의 공간에 순환을 접목시켰다.

USING FOREST RESOURCES THAT ARE ALREADY THERE

역성장 사회의 합리성: "가지고 있는 것을 최대한 활용한다"

역성장 사회의 합리성인 "초소형-주택 양식"으로 다음 나열된 가치를 재평가하여 지역적 특수성을 지닌 공간으로 재구성했다.

"가지고 있는 것을 최대한 활용한다."

산림자원이 풍부한 히다 지역에서 이미 주변에 존재하는 것으로부터 새로운 가치를 찾을 수 있는 가능성을 제안했다. "히다 지역에서 이미 주변에 존재하는 것으로부터 새로운 가치를 찾는다"는 아이디어를 바탕으로 네 가지 시각에서 현지 나무를 적절하게 활용하는 방법에 대한 그 가능성을 제안했다.

1. 삼나무 간벌재 활용
_ 작고 지름이 짧은 나무에서 나온 삼나무 간벌재는 건축용으로 사용된다."
간벌한 목재는 급경사지에서의 간벌 작업으로 인해 운반이 용이하도록 2m 정도로 절단한다. 이러한 이유로 삼나무 간벌재는 건축자재로는 부적합하지만, "목재 공법"을 적용하고, 소구경 목재와 긴 나사를 일체화한 "라미네이트 패널 공법"을 채택하여 길이가 짧은 소구경 삼나무 간벌재를 건축자재로 전환하려는 움직임이 있다. 강도가 다른 작고 불안정한 삼나무 목재를 적층하고 연결하는 작업을 통해 구조물의 전체적인 중복성을 늘리는 동시에 더 크게 만들 수 있었고, 가구 및 부속품에 사용할 수 있었다. 또한 각 목재의 이음매 간격을 조정하고 구멍을 만들어 변화하는 빛이 들어갈 수 있는 깊이와 틈이 있는 공간을 만들었다. 공법상 특별한 기술이 필요하지 않기 때문에 간벌에서부터 시공 과정까지 많은 사람들이 참여할 수 있었다.

2. 단단한 간벌 목재
- 필요시 간벌재의 유용성을 가치로 환산하는 방법"
나무를 최대한 살리는 조명. 각각의 조명은 다른 종류의 나무로 만들었다. 나무 줄기 자체의 형태와 나무껍질의 질감을 살려 디자인하였고, 가구 장인들이 홈을 파고 LED를 내장해 조명을 만들었다. 나무 베기에서 시작되는 식탁. 식탁은 나무 간벌

프로젝트 수행을 위해 모아둔 목련나무를 베는 것에서부터 시작된다.

3. 폐목재 사용
삼나무 폐목재로 만든 커피 테이블. 단일 통나무에서 가구와 건축자재로 쓰이는 목재는 약 25%에 불과하고, 나머지는 목재 건조용 연료로 쓰인다. 나머지는 목재 건조용 연료로 쓰인다. 폐목재 다발을 원기둥 모양으로 만든 커피 테이블은 공방에서 각 유닛을 묶어 만들었다.

4. 수선
중고 가구의 장기 사용을 위한 수선 및 복원. 식당 의자 8개 중 4개는 히다산교공업(Hidasangyo Industries)이 한때 판매했던 중고 '맥킨리' 의자를 수리해 재활용한 것이다. 이 의자들은 기후현 야마가타시의 감 타닌으로 다시 칠해 시간이 지날 수록 변하는 색감을 즐길 수 있다.

From Culture Consuming Tourism to Culture Producing Tourism

This project is a small hotel located in Hida Takayama, a flourishing tourist destination that attracts about 4 million visitors a year and is home to the World Heritage Site of Shirakawa-go and old streets. Here, while coexisting with conventional tourism that traces historical facts, we pursued the idea of the hotel as a base for generating culture that nurtures the future, making Hida Takayama a starting point for living regional values that will continue into the next 100 years of culture.

How do we create regionalism in a sustainable way after modernization?

About 100 years have passed since the establishment of modernist architecture. Today, with the benefits of modernization, we are able to enjoy services and lifestyles that are as close to homogenous as possible, regardless of whether we live in a city or a rural area. However, buildings and spaces made of industrialized materials are erasing the uniqueness of each region. When we think about the next 100 years, we would like to derive the uniqueness of the city in a sustainable way, after accepting modernization, so that the way we live now and interact with the environment itself can be passed on with value as a legacy to the future.

Micro-Vanacular" to Create Indigenousness

Thinking about what architecture can do to reconstruct the indigenousness of regions that have been homogenized by modernism for the next generation, I began by reconstructing the smallest units that make up these regions with things that already exist within a short distance. We called the subdivision of the indigenous vanaculer to a scale that can be realized in the shortest distance "micro vanaculer," and aimed to create a new form of architecture that can be successfully incorporated into the larger framework of social systems, including the economy and law, and constructed from within the environment and social cycles. The architectural act is to "make the most of what

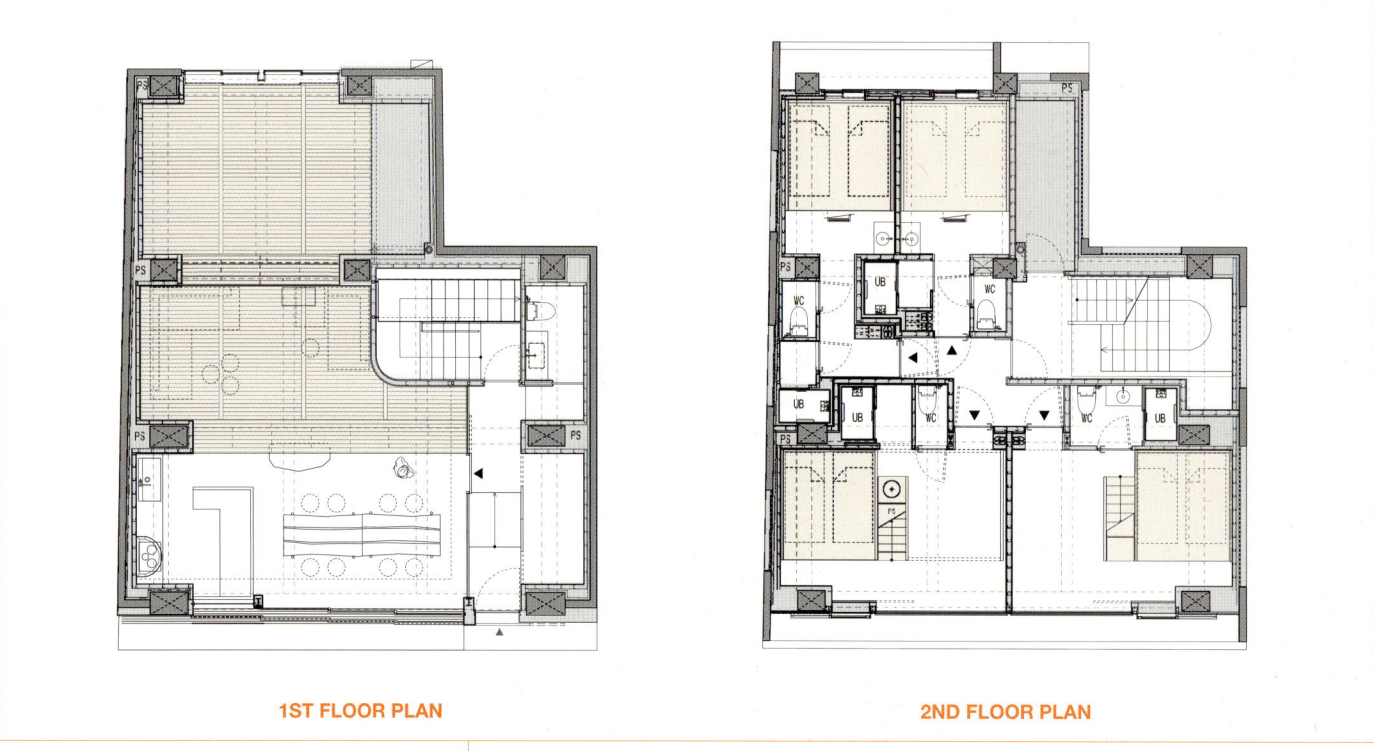

1ST FLOOR PLAN **2ND FLOOR PLAN**

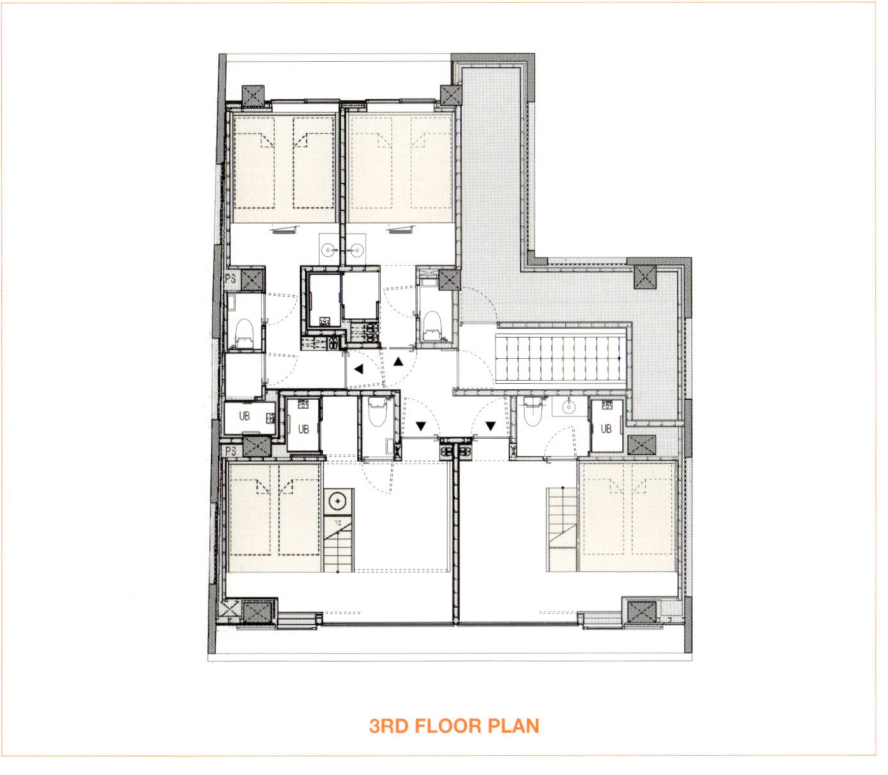

3RD FLOOR PLAN

is there," and by connecting it to the culture of living with the abundance of forests, we have incorporated circulation into the space in the shortest possible distance.

The Rationality of a Degrowth Society: "Making the Most of What We Have"
By reevaluating the values listed below as the rationality of a de-growth society: "micro vanacula", we have reconstructed them as a space with regional specificity.

"Make the most of what you have."
In the Hida region, which is rich in forest resources, we proposed the possibility of finding new value from what is already around us. Based on the idea of "finding new value from what is already around us in the Hida region," we proposed the possibility of utilizing the trees in the region in an appropriate manner from four different angles.

1. Utilization of thinned wood from cedar trees _ Cedar thinnings from short, small diameter trees are used for architectural purposes. The thinned wood is cut to about 2 meters for easy transportation due to the thinning work on steep slopes. For this reason, cedar thinning is unsuitable as a building material, but we are attempting to turn short, small-diameter cedar thinnings into a building material by applying the "log construction method" and adopting the "laminated panel construction method," which integrates small-diameter lumber with long screws. By laminating and connecting small, unstable cedar lumber with varying strength, we were able to increase the overall redundancy of the structure while making it larger, and we were able to use it for furniture and fittings. In addition, by adjusting the intervals between the joints of each piece of wood and creating a cavity, we were able to create a space with depth and a gap for the changing light to enter. In addition, since the construction method does not require

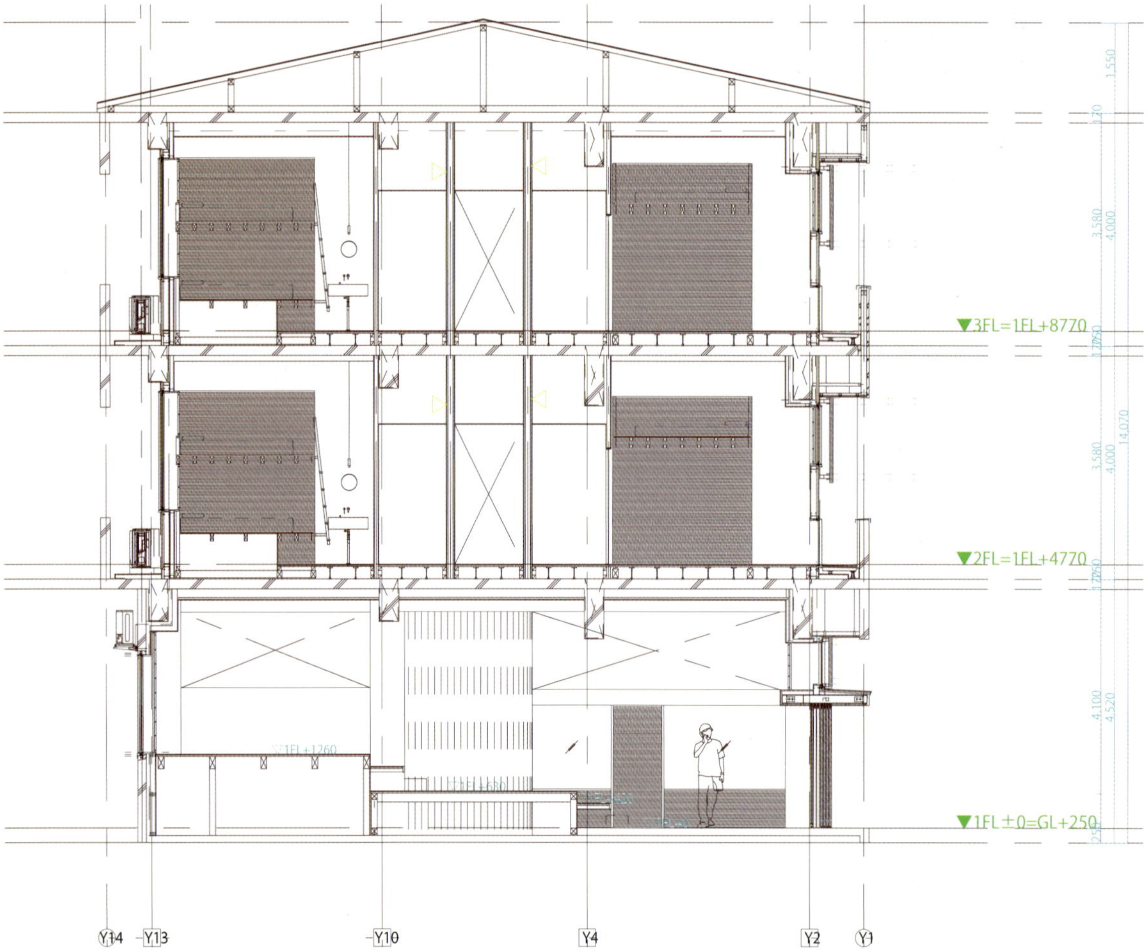

SECTION DETAIL

Neighborhood Facility

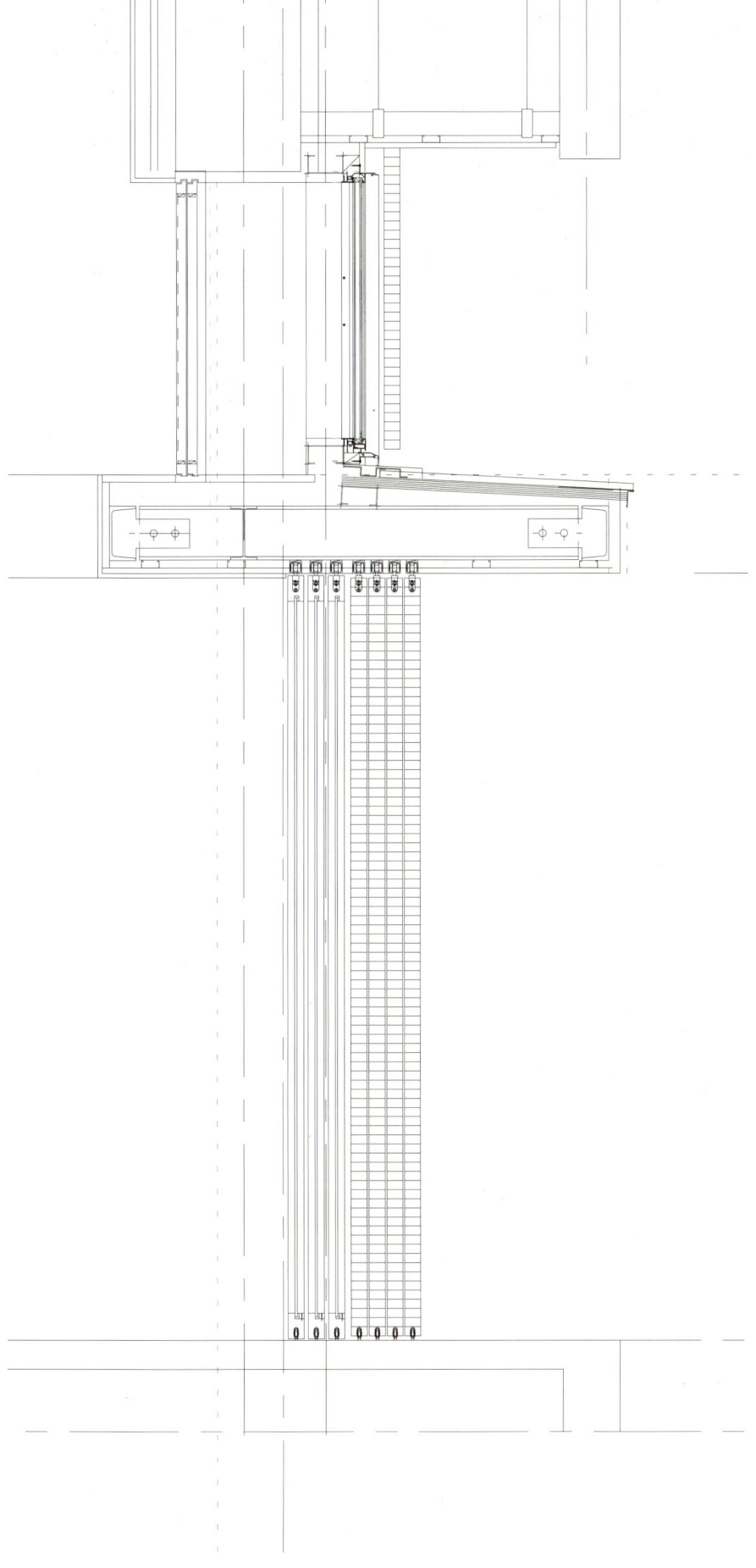

DETAIL

any special skills, many people are able to participate in the process from thinning to construction.

2. Hardwood thinned wood
-How to convert the value of thinned wood into value, if necessary.
Lighting that brings out the best in trees Each of the lights is made of a different species of tree, and is designed to take advantage of the shape of the trunk itself and the texture of the bark, while furniture craftsmen cut grooves and embedded LEDs to create the lighting. A dining table started from a tree cutting A dining table started from a tree cutting, using a magnolia tree that the members collected when they accompanied a tree thinning project.

3. Use of scrap wood
Coffee table made from cedar scrap From a single log, only about 25% of the wood is used for furniture and building materials, and the rest is used as fuel for drying the wood. The rest is used as fuel for drying the wood. The coffee table made of bundled scrap wood in the shape of a cylinder was made by bundling each unit in a workshop.

4. Repair
Repairing and Reviving Used Furniture for Long-Term Use Four of the eight dining chairs have been repaired and reclaimed from used "McKinley" chairs that were once sold by Hidasangyo Industries. They have been repainted with persimmon tannin from Yamagata City, Gifu Prefecture, and the color changes over time can be enjoyed.

Can be made larger by accumulating short materials.

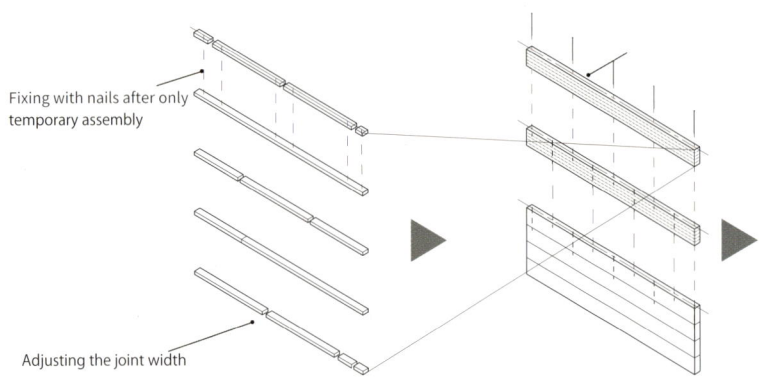
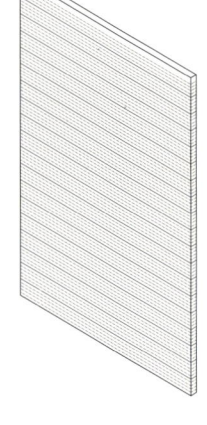

Fixing with nails after only temporary assembly

Adjusting the joint width

Small materials are temporarily assembled at the processing plant and brought to the site.

Staggered long screws are used on site to connect and integrate each material.

Since it is a combination of simple operations, anyone can install it anywhere.

Large furniture made of integrated panels

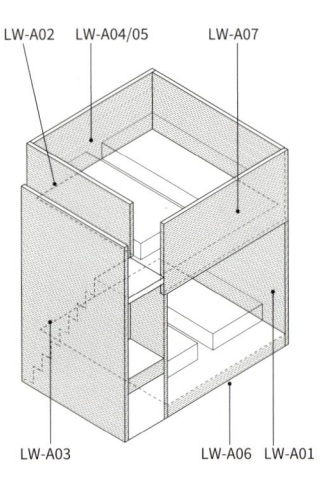
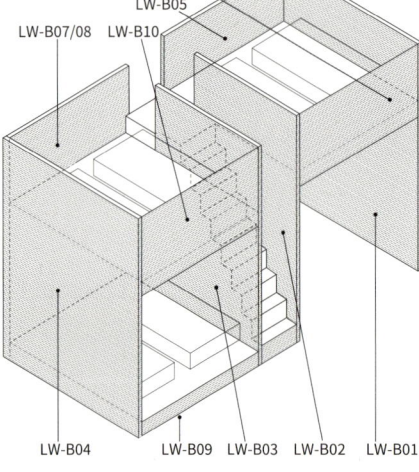
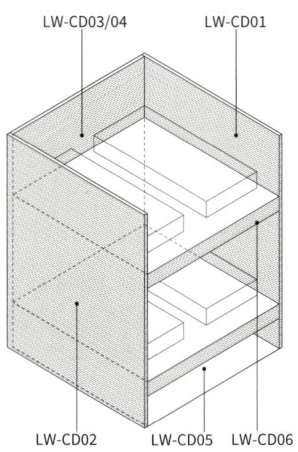

Guest room A

Guest room B

Guest room C·D

UTILIZATION OF THINNED WOOD FROM CEDAR TREES

Atsushi Nakamura / Kraft Architects

'We learn from the wisdom of our predecessors that remains in the present to create new spaces that renew our times.'
Kraft Architects is architectural design office based in Tokyo, Japan. The firm specialises in the design of houses, villas, resorts, hotels, inns, hot springs and public bathhouses, and also undertakes planning in a wide range of fields from architecture to interior design, landscape design and town planning.

Brief biography
2022 Part-time lecturer, Nihon University
2021 Part-time lecturer, Chiba Institute of Technology
2016 Kraft Architects established
2010 Tetsuo Furuichi Institute of Urban Architecture
2010 Graduated from Chiba Institute of Technology (Master's Program)
1985 Born in Takayama City, Gifu Prefecture

Awards
2023 German Design Award 2023 Gold Prize
2022 iF DESIGN AWARD 2022 Winner
      Wood Design Award 2021 Jury President's Prize
      The Architecture MasterPrize 2021 Winner
      Kukan Design Award 2021 Sustainable Award
2020 SKY DESIGN AWARDS 2020 Selected
      The Architecture MasterPrize 2020, Honourable mention
2019 Architects of the Year 2019 (The Architectural Design Association of Nippon)
2018 Wood Design Award 2018 Honourable Mention

Location
Gangnam-gu, Seoul, Republic of Korea
Use
Commercial, office
Site area
227.30m²
Built area
113.41m²
Total floor area
729.86m²
Floors
B1, 6F

Exterior finish
White brick, big slab ceramic tile
Interior finish
Exposed concrete
Project architect
Lee Jong-hee
Construction
C&O enc
Photographer
Kim Chang-mook

GOMBI

곰비빌딩

㈜오엠엠건축사사무소
omm architects

GOMBI는 서울 강남구 삼성동의 조용한 주택지 내에 위치한 근린생활시설 건물이다. 지하 1층, 지상 6층, 총 주차대수 5대 규모의 꼬마빌딩으로 게임, IT 등 소규모의 창의적인 회사를 위한 사옥 임대목적으로 계획되었다. 건축주는 업무공간으로서 기능적이고 효율적이며 요란하지 않게, 그러나 창의적인 회사를 위해 어느 정도 개성 있는 건물이 되길 원했으며, 지상 4층 정도부터는 서측으로 선릉뷰를 누릴 수 있는 높이가 되는 곳이라 발코니 등의 외부 휴게공간이 있으면 좋겠다는 요청을 하였다.

대지 규모가 작은데 반에 3종 일반주거지역이라 건폐율이 낮고 용적률이 높아 최대 용적을 위해서는 건물이 높이 올라가게 될 수밖에 없는데 일조권 사선제한에 상당부분이 영향을 받는 조건이었다. 일조사선의 영향으로 억지스러운 형태가 되지 않도록 최대 용적을 유지하면서 서랍처럼 매스를 넣었다 뺐다 엇갈리게 하면서 사선에 맞춰 자연스럽게

상층부로 올라가면서 후퇴하는 방식으로 매스를 디자인 하였다. 이렇게 함으로서 자연스럽게 선릉뷰를 누릴 수 있는 야외 테라스도 형성이 되었다.

건물의 전면으로 선큰을 만들고 1층 바닥을 그라운드 레벨에서 1.5m 들어올리고 1층에서부터 연속된 커튼월 전면유리를 지하층 선큰까지 하나의 입면으로 디자인하여 지하층이지만 풍부한 자연채광을 확보하였다. 거리에서 커다란 유리를 통해 지하층 바닥까지 시선이 닿게 되어 지하층이지만 충분한 개방감으로 골목길과 소통면서 지하층의 상업성을 높일 수 있었다. 지하층의 후면으로도 작은 선큰을 계획하여 양방향 피난문제를 해결하고, 전후면 선큰을 통해 맞바람이 통하도록 하여 쾌적한 지하공간이 되도록 하였다. 동측에는 유리로 시원하게 개방된 직통계단이 옥상정원까지 연결되며, 옥상정원은 가벽과 캐노피로 아늑한 작은 휴게공간이 되어 선릉풍경을 보면서 회사의 다양한 이벤트공간으로 활용될 수 있도록 하였다.

외장재는 백고벽돌을 사용하여 차분하게 주택가 동네에 어울리도록 하였고, 조형적으로 유니크하게 들어올려진 1층의 외부면에는 백고벽돌로 된 전체 외관과 확연하게 상반된 느낌의 외장재로 입구성을 강조하고자 하였다. 작은 스케일의 벽돌과 달리 1,000x3,000 규격의 빅슬래브 박판타일을 사용하고, 색깔 역시 아이보리색의 벽돌과 또렷이 대비될 수 있는 붉은 코르텐 색상의 타일을 적용하여 전체건물의 차분하고 심플한 느낌과 완전히 상반되는 저층부의 모습을 연출하였다.

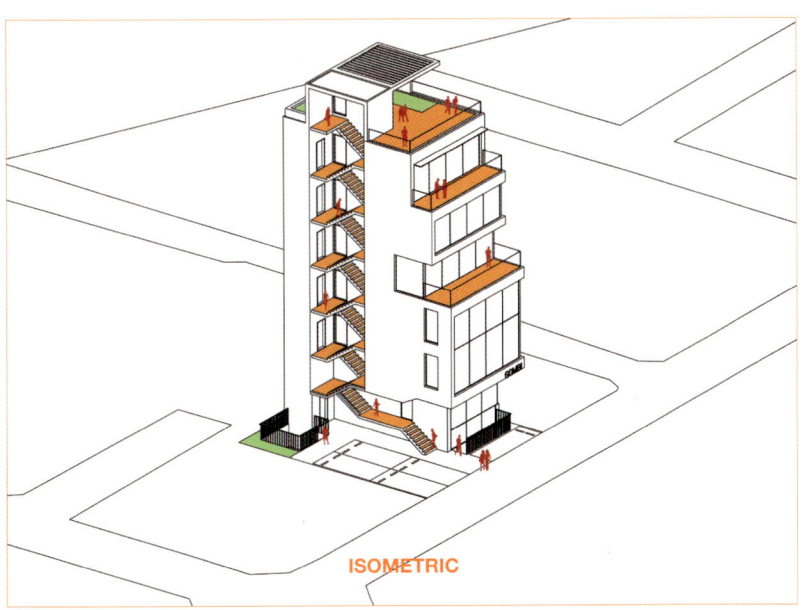

ISOMETRIC

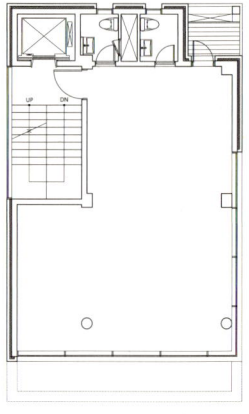
5TH FLOOR PLAN

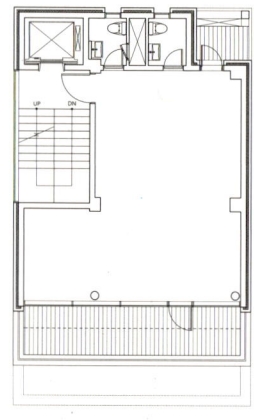
6TH FLOOR PLAN

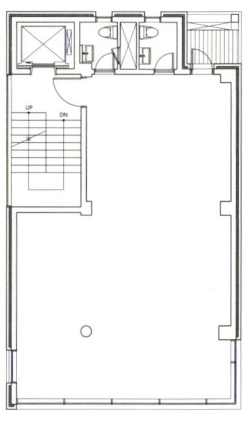
2ND FLOOR PLAN

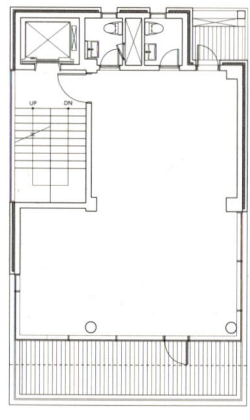
4TH FLOOR PLAN

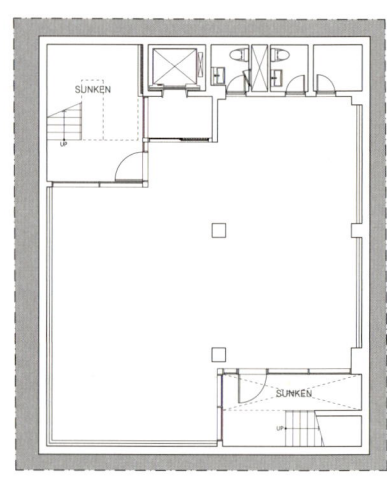
B1 FLOOR PLAN

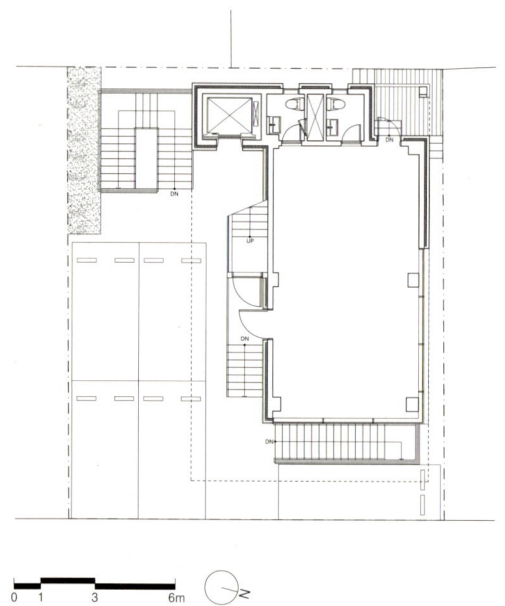
1ST FLOOR PLAN

Neighborhood Facility

GOMBI is a neighborhood living facility situated in a quiet residential area in Samseong-dong, Gangnam-gu, Seoul. It is a small structure with one basement floor and six stories above ground, and a total of five parking spaces. The client wanted the building to be utilitarian, efficient, and unobtrusive as a workspace, but to have some personality for a creative company, and requested outdoor rest areas such as balconies from the fourth floor, since you can enjoy the view of Seolleung to the west from that height.

Although the size of the land was small, it was a class 3 general residential area, so the building-to-land ratio was low, and the floor area ratio was high. Therefore, the building had to be raised high for maximum volume, but it was largely affected by the architectural slant line for daylight. The masses were designed to resemble drawers, pushed and pulled to avoid awkward shaping due to the restrictions of the daylight slant line, while still maintaining maximum volume. In accordance with the slant line, they recede more at higher parts. In this way, outdoor terraces, where you can enjoy the view of Seolleung, were also formed.

A sunken area was created at the front of the building, and the first floor was raised 1.5 meters above the ground level. The front curtain wall glass was designed to be a single façade extended from the first floor to the sunken area on the basement floor, letting abundant natural light reach all the way to the basement floor. The large glass walls, allowing the view from the street to reach the basement floor, increased the commercial value of the basement floor as it opened to the outside street. Another small sunken area was planned at the rear side of the basement floor to address the

issue of emergency evacuation in both directions and encourage cross winds to flow through the sunken sections on both sides to create pleasant underground spaces. A glass-encased stairway on the eastern side, with unhampered openness, leads directly to the rooftop garden, which is a quaint little resting spot with makeshift walls and canopies that can be used as a space for various corporate functions with an enjoyable view of the scenery of Seolleung.

Off-white bricks were used for the exterior to subtly complement the residential neighborhood. The façade of the ground floor was raised in a distinctively formative manner, emphasizing the entry using exterior materials that clearly stood out against the overall off-white brick exterior. Big slab thin-plate tiles in the 1000 x 3000 size, in contrast to the small-size bricks, as well as reddish corten steel-colored tiles, which contrast sharply with the ivory-colored bricks, were employed to give the lower levels a sense that is completely opposite to the unassuming and straightforward feeling of the entire building.

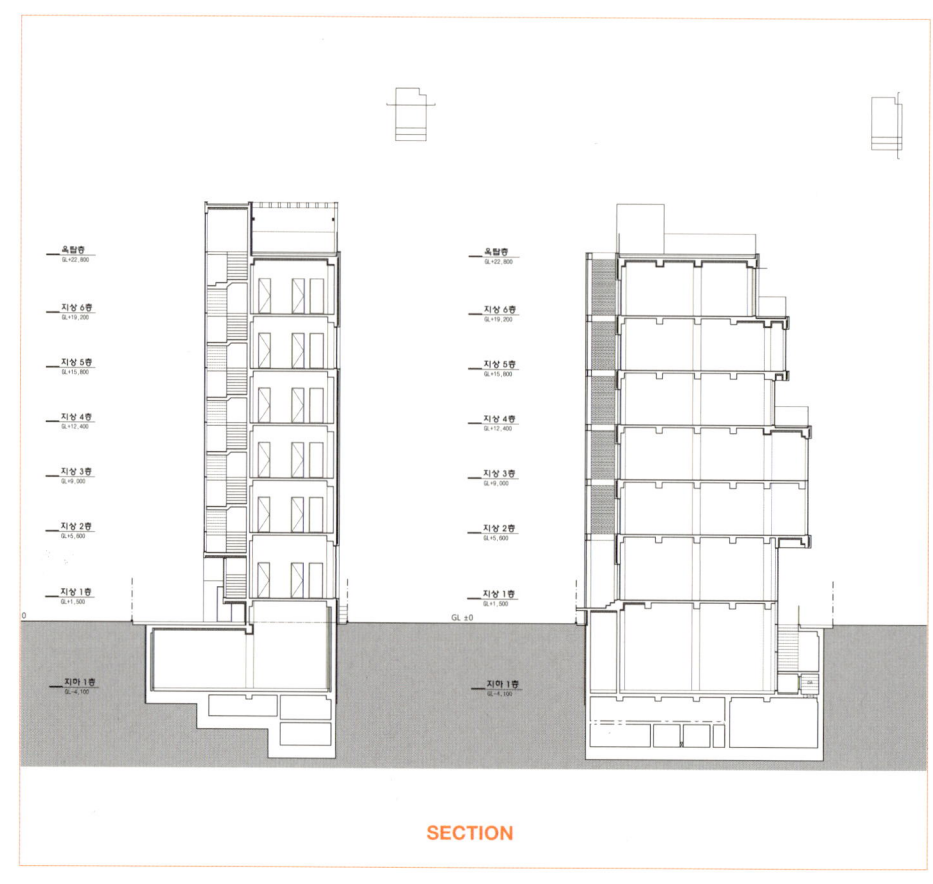

SECTION

Neighborhood Facility

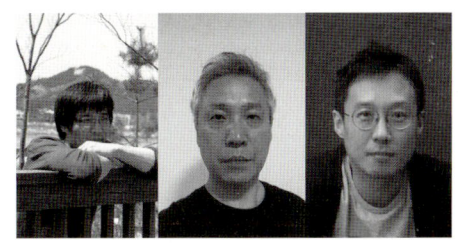

김원영, 박남규, 이종희 / (주)오엠엠건축사사무소
Kim Wonyoung, Park Namkyu, Lee Jonghee /
omm architects

오랜 시간 공간건축에서 근무해온 3명의 소장이 함께
건축설계사무소 topos를 창립으며, 2023년 omm으로
사명을 변경했다.

오엠엠건축사사무소 대표 건축가들은 각 13년~20년의
시간동안 공간건축에서 주택, 근린생활시설부터
대규모 공공턴키, 현상설계까지 충분히 다양한 경험을
축적하였다. 이들은 이 기간을 새로운 건축을 하기 위한
수련의 과정이었다고 생각한다.

건축은 이미 거장의 시대를 넘어 협력과 통섭을 기반으로
새롭게 성장하고 있다. 따라서 omm은 한 사람의 독자적
아이디어에 따르기보다는 다양한 사람, 지식, 문화, 기술
분야와의 협력과 수평적 조직을 지향한다. omm은 보다
새롭고 진보적인 도시의 공공성과 건축주의 의도에
기여하는 것을 목표로 하고 있다.

Neighborhood Facility

Location
Beijing, China
Use
Hotel
Site area
250m²
Built area
120m²
Total floor area
249m²
Floors
2F

Exterior finish
Coating, concrete
Interior finish
Latex paint, concrete, wood
Project architect
Le Sheng, Wei Wang
Photographer
Atelier d'More

Sleeping Lab·Arch

슬리핑 랩-아치

Atelier d'More

Neighborhood Facility

BEFORE

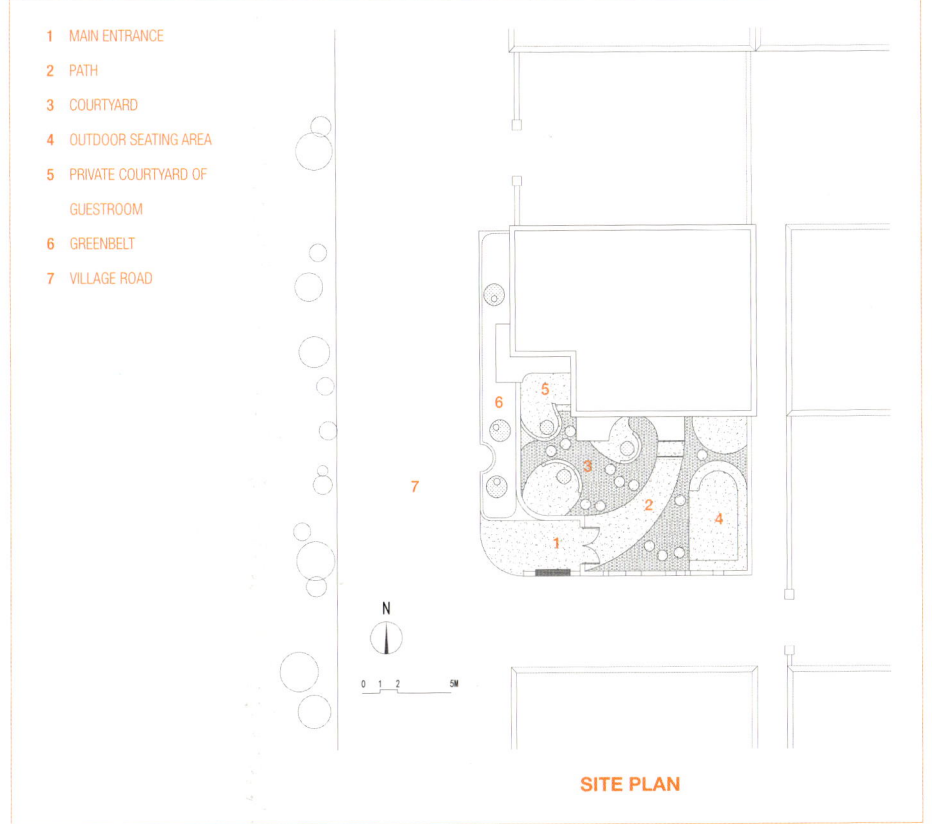

1 MAIN ENTRANCE
2 PATH
3 COURTYARD
4 OUTDOOR SEATING AREA
5 PRIVATE COURTYARD OF GUESTROOM
6 GREENBELT
7 VILLAGE ROAD

SITE PLAN

Project overview

The project is located in Huangmuchang Village on the outskirts of Beijing. It is a two-story brick-concrete building, originally used as residential and office. The weather-beaten exterior walls and wind-blown eaves are eager to be refurbished and restored. Atelier d'More was commissioned to renovate this run-down building into a boutique hotel.

The architectural intention of creating a landscape

Given the lack of scenery in the surrounding environment, we decided to create an inward view and create an interesting private garden, which led to the idea of the enclosed courtyard space. Only a few viewing frame are opened on the south wall. In addition to the blue sky and white clouds, the chaos and noise are isolated from the wall. The large glass windows of the building actively introduce the scenery and light into the building, which becomes a part of the indoor space experience.

We divided the rectangular courtyard of 100 square meters into several small semi-enclosed courtyards, trying to create an inner courtyard environment. Each courtyard will have a small tree as the protagonist (since it is completed in winter, the tree has not been planted), an organic connection is formed between the courtyards. A series of interactions between tearoom, guest rooms and the courtyard are created under the multi-level space. When users enter the building, they can read the interior of the building in their own way and perspective, so the container function of the building is reflected. The shadow moves with the light, and the scenery changes with people, making the architectural space full of poetry.

Free flow

The design language of the architectural and interior is based on a quarter arc,

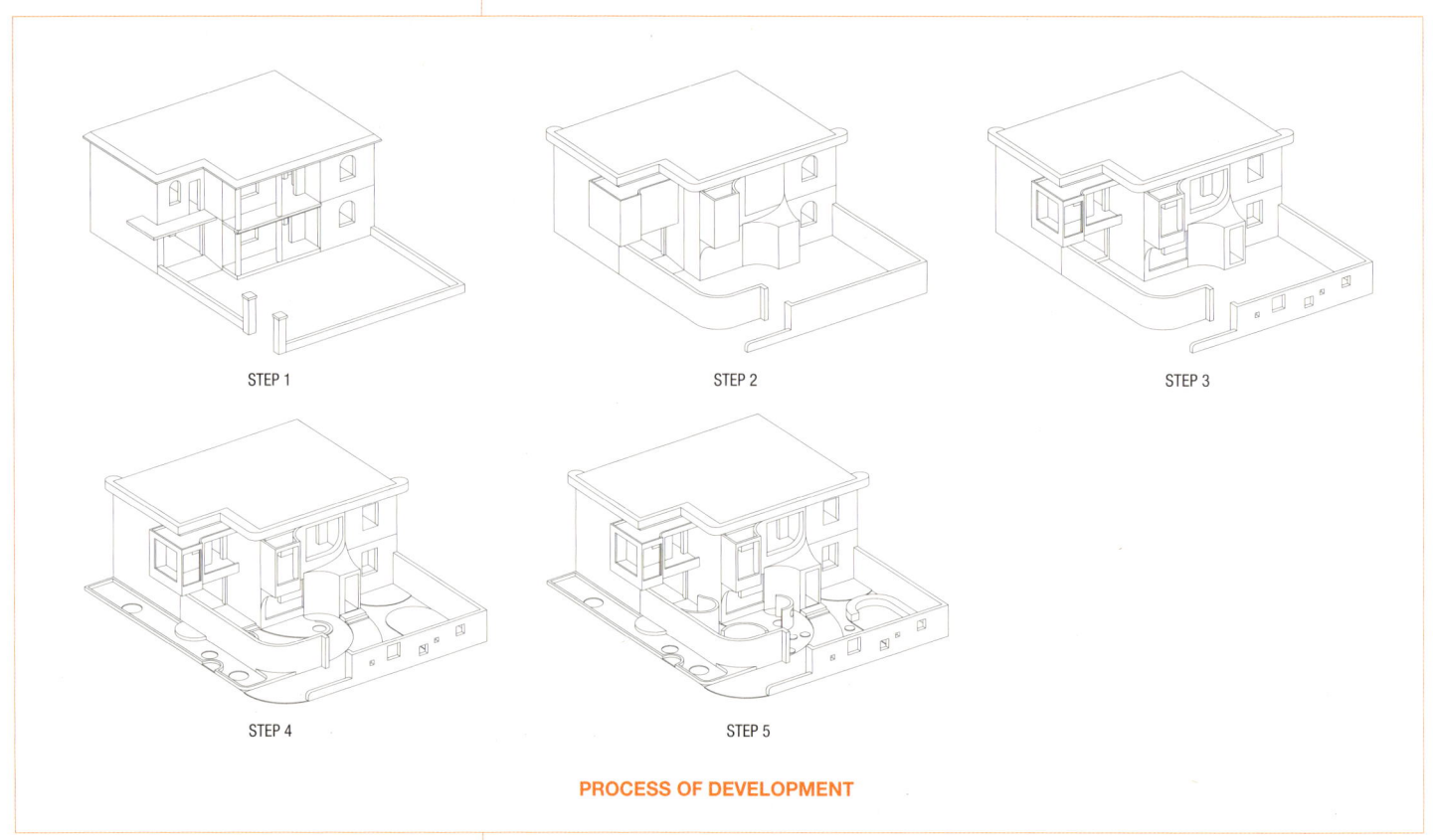

STEP 1　　　　　　　　STEP 2　　　　　　　　STEP 3

STEP 4　　　　　　　　STEP 5

PROCESS OF DEVELOPMENT

Neighborhood Facility

which flows freely in the space with a rigorous attitude and increases the tension of the building by means of geometric composition. The outward extension of the building block on the facade is close to nature in a flowing posture, bringing the distance between architecture, nature and people closer. The white facade makes the building abstract and sculptural, through the rendering of white material, the boundary between the wall and the top is blurred. The flowing space and natural light blend together to bring people a pure experience and imagination space.

Assembling "Toy"

All the furniture in this project were designed by Atelier d'More, in fact, we have been engaged in the designing prefabricated assembled furniture for many years, different from the manufacturing industry, the traditional decoration industry automation process develops slowly, our concept of furniture assembly is equivalent of turning on automatic gears for construction, that the clients can even operate themselves. We regard this kind of assembling furniture as a big toy, making the boring construction work become a game, in which the workers and clients can enjoy the process. Many of the interior decorations can be assembled on-site with prefabricated components, which has been implemented in many of our projects. The application of this design method is very meaningful for low construction cost projects.

프로젝트 개요

본 프로젝트는 베이징 외곽의 황무창 마을에 위치해 있다. 2층 벽돌 콘크리트 건물로, 기존에 주거지와 사무실로 사용되었던 2층 벽돌 콘크리트 건물은 햇볕에 거칠어진 외벽과 바람에 날린 처마는 개조와 복원이 절실했다. 아틀리에 디모어(Atelier d'More)는 이 낡은 빌딩을 부티크 호텔로 개조하는 일을 의뢰 받았다.

경관을 조성하려는 건축적 의도

주변 환경의 경관이 부족하다는 점을 감안하여 내부 전망을 만들고 흥미로운 개인 정원을 만들기로 결정했으며, 이는 폐쇄된 중정 공간이라는 아이디어로 이어졌다. 남쪽 벽에는 몇 개의 조망용 프레임만 열려있다. 파란 하늘과 흰 구름 외에도 혼돈과 소음이 벽으로부터 격리되어 있다. 건물의 대형 유리창은 건물 안으로 풍경과 빛을 적극적으로 유입시켜 실내 공간 체험의 한 부분이 된다.

100m² 직사각형 중정을 여러 개의 작은 반폐쇄 중정으로 나누어 내부 중정 환경을 조성하고자 했다. 각 중정마다 작은 나무를 주인공으로 하여(겨울에 완성해 나무가 심어져 있지 않음) 중정 간에 유기적인 연결이 형성될 것이다. 다층 공간 아래에서 다실, 객실, 중정 사이의 일련의 상호 작용이 생성된다. 사용자가 건물에 들어서면 자신만의 방식과 관점으로 건물 내부를 읽을 수 있어 건물의 컨테이너 기능이 반영된다. 그림자는 빛과 함께 움직이고 풍경은 사람과 함께 변화하여 건축 공간을 시로 가득 채운다.

자유로운 흐름

건축과 인테리어의 디자인 언어는 4분의 1 호(arc)를 기반으로 엄격한 태도로 공간을 자유롭게 흐르고 기하학적 구성을 통해 건물의 긴장감을 높인다. 정면부의 건물 블록을 바깥으로 확장한 것은 흐르는 자세로서 자연과 가깝고 건축과 자연, 사람 사이의 거리를 더 가깝게 만든다. 하얀 정면부는 건물을 추상적이고 조각처럼 보이게 만들며, 흰색 재료를 렌더링하여 벽면과 상단 사이의 경계가 흐려진다. 흐르는 공간과 자연광이 어우러져

Neighborhood Facility

1 FOYER
2 RECEPTION
3 TEA ROOM
4 DINING AREA
5 OPEN KITCHEN
6 STAFF ROOM
7 STANDARD ROOM
8 FAMILY SUITE ROOM
9 CORRIDOR
10 SUITE ROOM WITH BALCONY

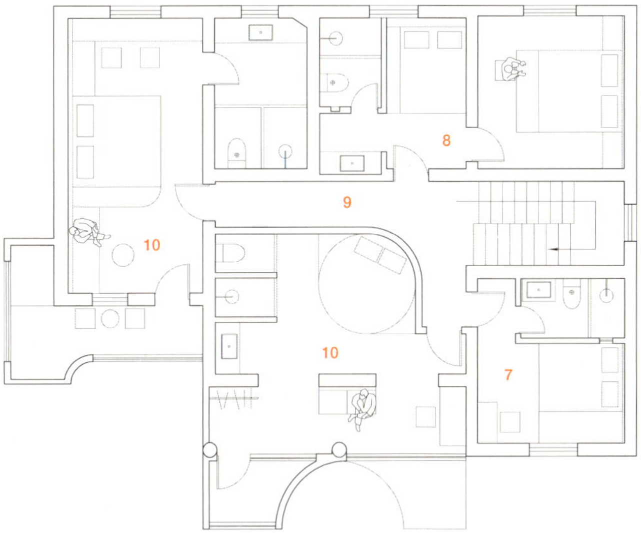

1ST FLOOR PLAN

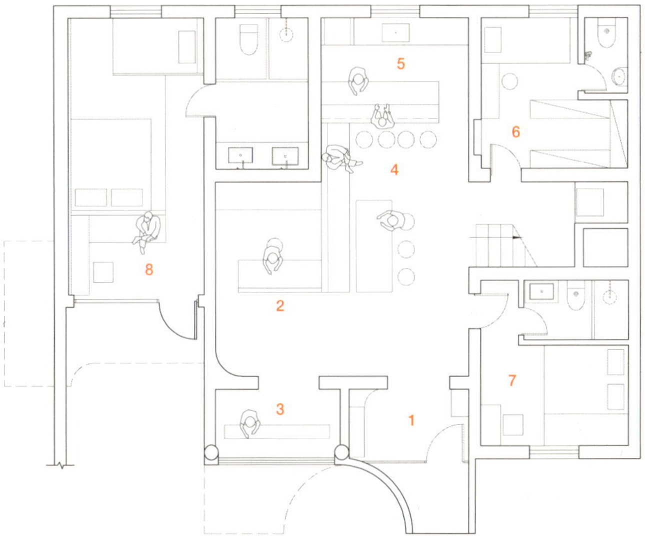

GROUND FLOOR PLAN

Neighborhood Facility

사람들에게 순수한 경험과 상상의 공간을 선사한다.

"장난감" 조립

이 프로젝트의 모든 가구는 아틀리에 디모어가 디자인했다. 사실 우리는 제조 산업과 달리 수년 동안 조립식 가구 디자인에 종사해 왔다. 전통적인 장식 산업 자동화 프로세스는 느리게 발전한다. 가구 조립의 개념은 건설을 위해 자동 기어를 켜는 것과 같으며 고객이 직접 작동할 수도 있다. 우리는 이런 조립식 가구를 하나의 큰 장난감으로 여기며, 지루한 시공작업을 작업자와 고객이 함께 즐길 수 있는 게임으로 만든다. 많은 실내 장식품들은 현장에서 조립할 수 있고, 조립품들은 우리의 많은 프로젝트에서 구현된 부품들이다. 이 설계 방법은 저비용 공사 프로젝트에서 그 의미가 크다.

Neighborhood Facility

235

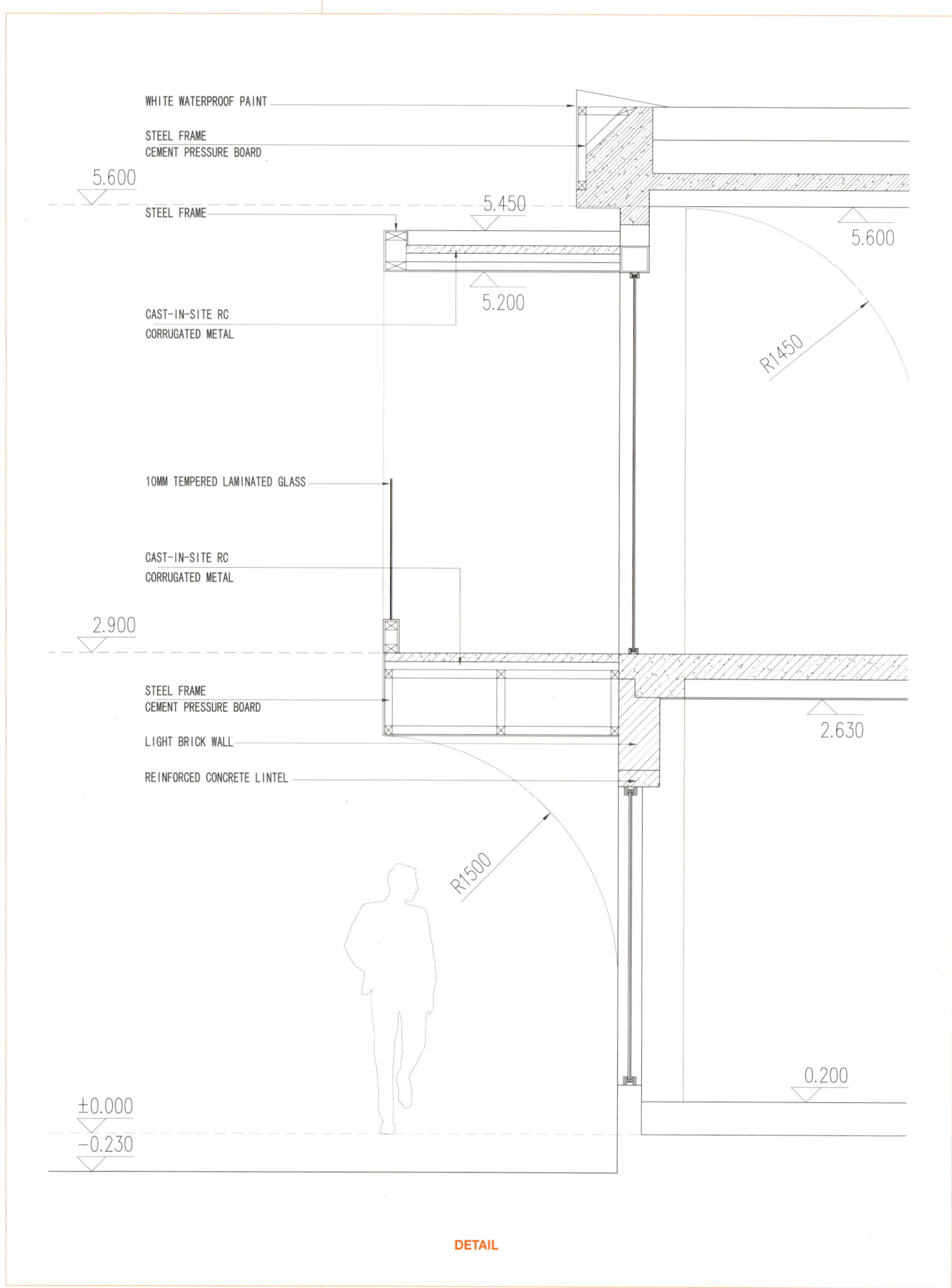

DETAIL

Neighborhood Facility

Atelier d'More

Atelier d' More was founded in Shanghai in 2015. It is a research-based design firm with innovation and practice as its core. The scope of our work covers architecture, interior and landscape design. The studio is committed to breaking traditional design aesthetics on the basis of architectural logic, creating surprises in the space with innovative and poetic expression. We always maintain the spirit of self-criticism in the design, pay attention to the real needs of space, so as to create unique design works that conform to the actual demand and rich in humanities.

>> www.d-more.cn

Neighborhood Facility

Location
Stockholm, Sweden
Use
Office
Site area
3,000m²
Built area
120m²
Total floor area
240m²
Floors
2F

Exterior finish
Glulam wood structure and Falu black painted wood boards in walls
Interior finish
Wall_ Pine wood plywood, Ceiling_ Pine wood boards, Floor_ Oak and spruce Boards
Project architect
Anders Berensson
Photographer
Anders Berensson

Outdoor Office

아웃도어 오피스

ANDERS BERENSSON ARCHITECTS

Neighborhood Facility

Anders Berensson Architects has designed a building that is half office, half outdoor meeting space in the Royal National City Park Norra Djurgården in Stockholm. The house is constructed as a framework of wood beams filled with wall panels and glass at the top part and left open at the bottom part creating a weather protected outdoor space.

The building is an extension of an existing office building inhabited by Swedish outdoor and shoe company Lundhags. Building in the royal national park we tried to do as little excavations as possible. Therefore, we decided to design a house on wood pillars standing on the sloping terrain.

The top part of the house follows the shape of the existing building with a slight tilt in direction to follow the existing landscape. This part is directed towards the forest looking into the crowns and trunks of the old trees creating a calm sensation. The bottom part is a concrete amphitheater casted on top of the existing terrain. Standing in a slope the bottom part is closed to the road in the north and opened towards the forest while sloping whit levels towards a stage for outdoor events.

Designing an office/outdoor structure in the forest we tried to find a balance between an office building whit a strict geometry and a traditional cabin. We decided to follow the shape of the existing building but design it as a wood framework. The primary structure is made of glulam beams and pillars, the wood frame is then filled with either glass windows, wood panel walls or left opened reinforced with wood crosses. All materials on the inside are made from wood, the ceiling is made from pine tree boards, the walls from pine plywood. The floor is made as an installation grid where electricity and data are installed under a spruce tree grid surrounding oak boards.

All details are kept simple either in wood or as black painted steel. Windows are milled into the wood pillars and beams, all lamps are cladded with the same wood as the ceiling they sit on. Most furniture's are made from leftover pieces from the building site. The tables are made from the same plywood sheets as the wall cladding. The lounge is made from left over glulam timber and other wood boards from the house. The chairs are cladded whit shoe leather from the Lundhags factory. All details that needed to be built in steel are painted black and, on the outside, sometimes shaped as animals from the royal national park. Some who lives there today and some that we hope one day will stroll back underneath the house.

앤더스 베렌슨 아키텍츠(Anders Berensson Architects)는 스톡홀름의 노라 유르고르덴 왕립 국립공원에 지어진, 사무실 겸 야외 회의 공간인 건물을 디자인했다. 이 건물은 상부는 벽판과 유리로 채워져 있고 하부는 개방된 채로 있는 목재 들보의 골조로 만들어져 있어 날씨의 영향을 받지 않는 야외 공간으로 형성되었다.

이 건물은 스웨덴의 아웃도어 및 신발 회사 룬드하그스(Lundhags)가 소재한 기존의 사무실 건물을 확장한 것이다. 왕실 국립공원에 건물을 지으면서 가능한 한 굴착을 적게 하려고 노력했다. 그래서 경사지에 서 있는 나무기둥 위에 설계하기로 결정했다.

집의 윗부분은 기존의 풍경을 따르기 위해 방향을 약간 기울여 기존 건물의 형태를 따른다. 이 부분은 오래된 나무의 관과 줄기를 들여다보며 숲을 향해 있어 차분한 느낌을 준다. 건물의 바닥 부분은 기존 지형 위에 콘크리트를 타설하여 만든 원형 극장이다. 경사지에 서 있는 바닥 부분은 북쪽의 도로를 향해서는 닫혀 있고 숲을 향해서는

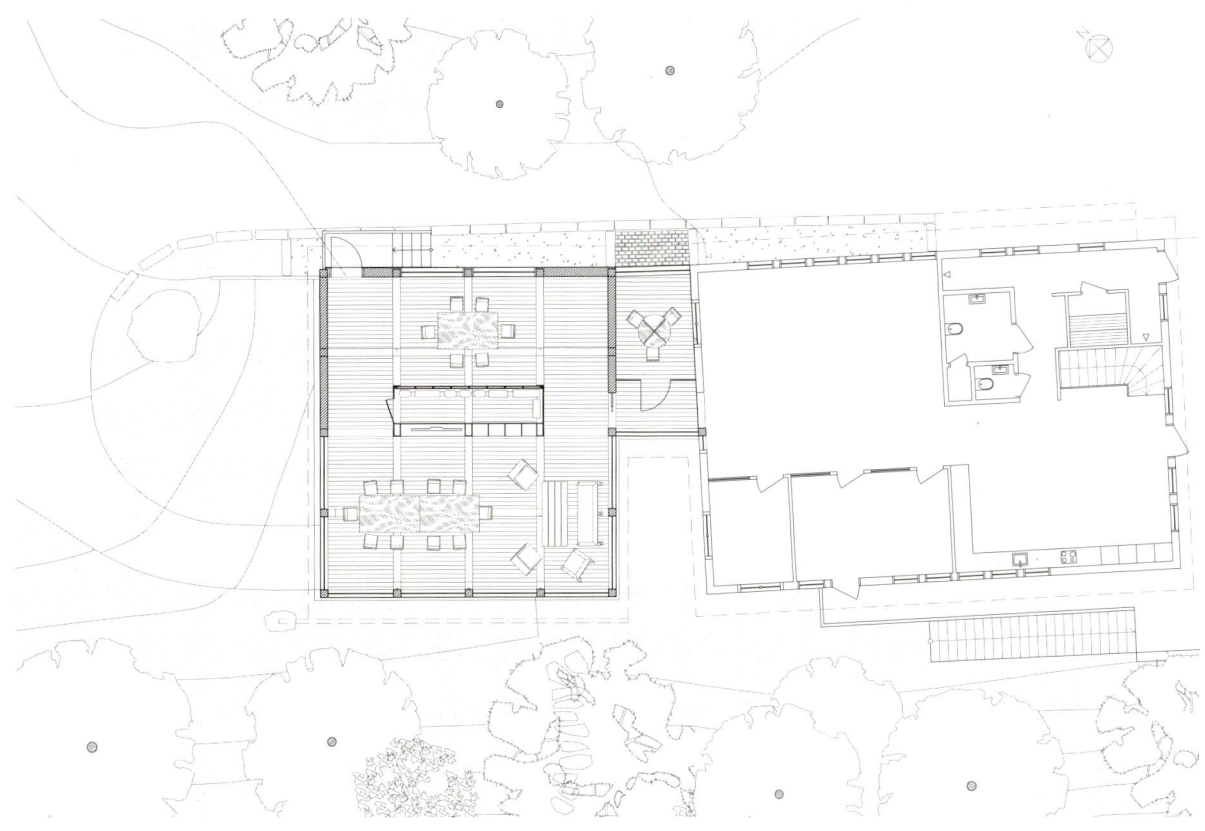

TOP FLOOR PLAN

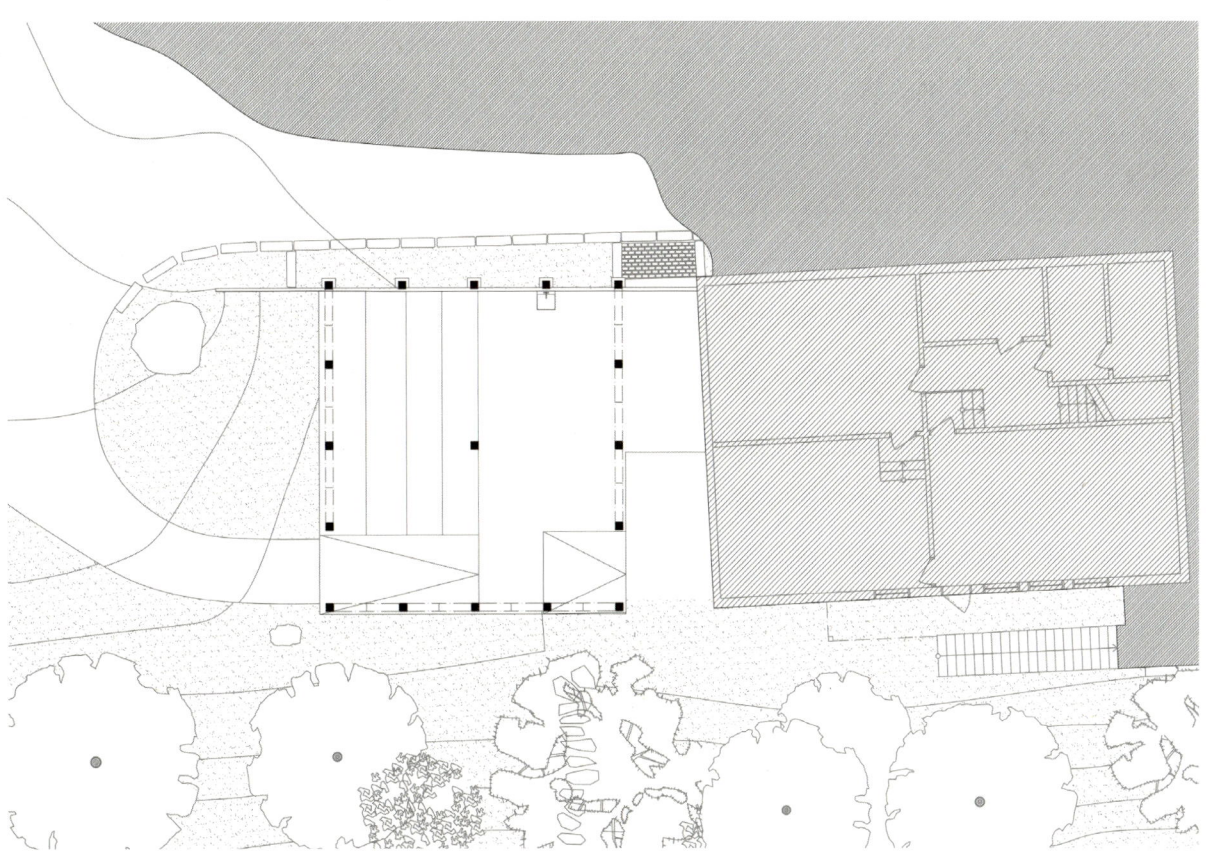

GROUND FLOOR PLAN

열려 있으며 야외 행사를 위한 무대 쪽으로 기울어져 있다.

숲 속 사무실·옥외 구조물을 디자인하면서 엄격한 기하학적 구조를 가진 사무실 건물과 전통적인 오두막 사이의 균형을 찾으려고 노력했다. 우리는 기존 건물의 형태를 따르되 목조 골조로 디자인하기로 결정했다. 기본 구조는 집성재 빔과 기둥으로 만들어지며 목재 프레임은 유리 창문, 목재 패널 벽으로 채워지거나 왼쪽으로 개방된 형태로 목재 십자가로 보강된다. 내부의 모든 재료는 목재이며 천장은 소나무 판자, 벽은 소나무 합판으로 구성했다. 바닥에는 참나무 판자를 둘러싼 가문비나무 그리드 아래에 전기와 데이터 전송을 위한 설치 그리드를 만들었다.

모든 디테일은 목재나 검게 칠한 강철로 심플하게 유지한다. 창문은 목재 기둥과 들보에 가공되어 있고, 모든 램프는 천장과 같은 목재로 덮여 있다. 대부분의 가구들은 건축 현장에서 남은 조각으로 만들었다. 테이블은 벽면 피복과 같은 합판 시트로 만들고, 라운지는 건물에서 남은 집성재 목재 및 기타 나무 판자로 만들었다. 의자는 룬드하그스 공장의 구두 가죽을 입혔다. 강철로 제작해야 하는 모든 디테일은 검은색으로 칠하고, 외부는 때때로 왕실 국립공원의 동물 모양을 하고 있다. 오늘날 그곳에 살고 있는 사람들도 있고, 언젠가 집 밑으로 돌아와 거닐기를 바라는 사람들도 있다.

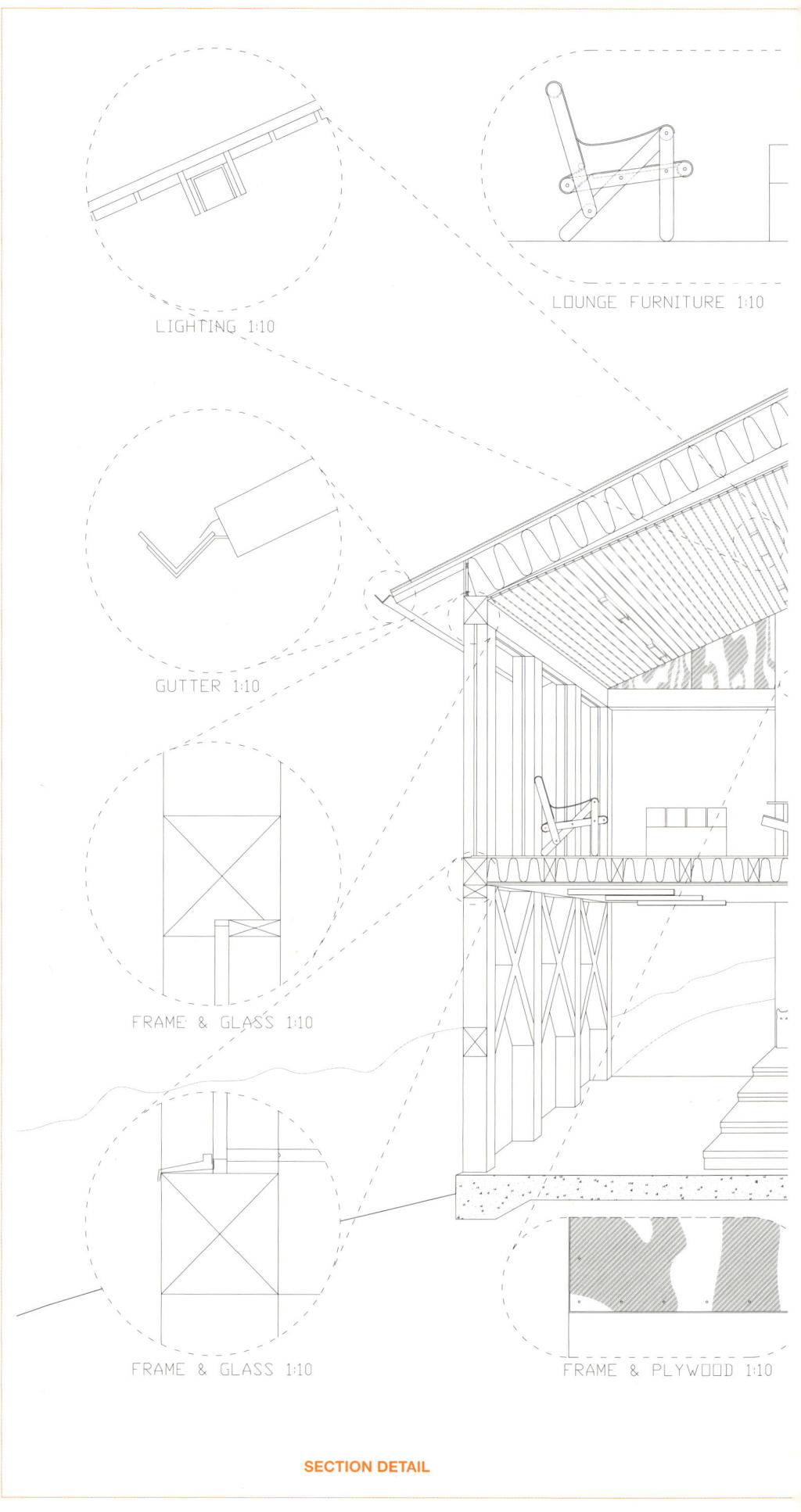

SECTION DETAIL

Neighborhood Facility

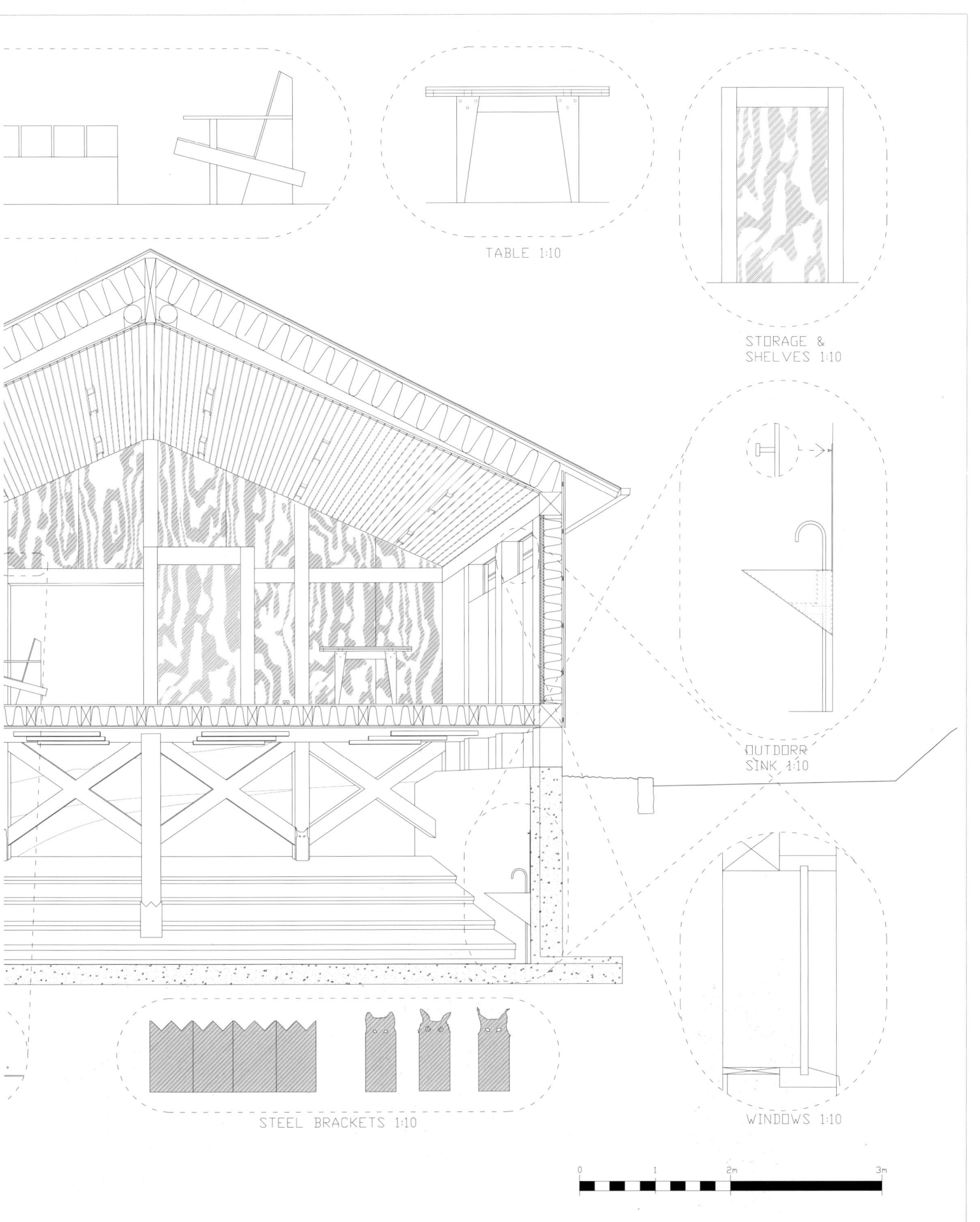

SECTION

Neighborhood Facility

Anders Berensson Architects

Anders Berensson Architects is an architecture office founded in Stockholm, Sweden by Anders Berensson in 2015. The philosophy of the office is to gain energy for each project from the unique conditions that each client and task provide. The office do not have a fixed design idea except that we only do great projects for great people that we, the client and society can be proud of. We work with all kind of clients, big or small. We like variety! The office have a consciously diverse portfolio since we have shifting clients with different needs and dreams from all over the world. Our reputation is not based on repeating ourselves but by inventing new great architecture everywhere for everyone. The firm has a portfolio that's been widely published all over the world and we can promise you a dedicated, unique and outstanding project.

\>> www.a-b-a.se

근린생활시설 7

121m² – 190m²

NEIGHBORHOOD FACILITY 7

THE FLYING WALLS HOSTEL / 더 플라잉 월 호스텔
KOMAL DHULIA / DHULIA ARCHITECTURE DESIGN

NGÓI SPACE / 타일 공간
H&P ARCHITECTS

KAMA-ASA SHOP / 가마아사(釜浅) 상점
KAMITOPEN CO., LTD.

SUNU BLDG. / 선우빌딩
엘케이에스에이 건축사사무소 LKSA

W 134 (INOSYS HEADQUARTERS OFFICE) / 더블유 일삼사(이노시스 사옥)
(주)오엠엠건축사사무소 omm architects

VIRA II / 비라 II
ALIDOOST AND PARTNERS

PENNANT THONGLOR / 레스토랑 페넌트 통로
PAD SPACE ARTISAN

Location
Rajkot, India
Use
Hostel
Site area
185.8m²
Built area
125.42m²
Total floor area
836.13m²
Floors
G+4F

Project architect
Komal Dhulia
Project Team
Kishan Makwana
Structural Engineer
Manish Doshi Office
Photographer
Dhrupad Shukla

The Flying Walls Hostel

더 플라잉 월 호스텔

KOMAL DHULIA / DHULIA ARCHITECTURE DESIGN

The flying walls hostel project was commissioned to us by a company intending to provide their workers with a living space that promotes community living and wellbeing. The site is located on the outskirts of the city of Rajkot in a developing area with educational, industrial and hospitality buildings coming up. Despite the development, the site's location provides unending views of the agricultural, grazing lands and the horizon.

The functional requirement was straightforward, the lower level of the building dedicated to the general managers and officials while the higher levels dedicated to other workers of different ranks. The company had 5 General Managers and approximately 50 other employees. The requirement divided the built mass into five equal levels with a simple floor plan and a site and services kind of approach. As the inhabitants hail from different parts of the country, a friendly and open environment in terms of built form was necessary to promote a healthy social bond amongst everyone.

The weather protection elements were primarily worked on to provide spaces that are ambient throughout the day. The heat from the sun after midday heats the building surfaces exposed to the west the most. To tackle this, we staggeringly stacked the walls above cantilevered slabs to create the necessary barrier.

This process sparked an idea to incorporate this in the building's form and language. What if we could cantilever the walls out of its resting slabs? As if the walls were flying out from the facade and balancing over the central frame structure. These flying walls develop an interesting range of shadows over the balconies creating a frame visually appealing. When pictured with the context a slight contrast is always maintained neither overpowering nor merging to create the

Neighborhood Facility

identity and incite a sense of pride and belonging in the occupants.

플라잉 월스 호스텔(Flying Walls Hostel) 프로젝트는 직원들에게 공동체 생활과 웰빙을 증진시키는 생활 공간을 제공하려는 회사로부터 의뢰 받은 작업이다. 이 건축 부지는 라즈콧시(Rajkot) 외곽에 위치하고 있으며 교육, 산업 및 접객 건물이 들어설 개발 지역이다. 개발에도 불구하고 건설 부지의 위치에서는 농업, 방목지, 지평선의 끝없는 전망을 볼 수 있다.

기능적 요구사항은 단순했으며, 건물의 하부층은 총괄 관리자와 임원에게, 상부층은 다른 직급의 직원들에게 할당되었다. 회사에는 총괄 관리자 5명과 직원은 약 50명이 있다. 요구사항은 간단한 평면도와 건설 부지 및 서비스 종류에 따라 건물을 다섯 개의 동일한 층으로 나누는 것이었다. 입주할 직원들이 모두 다른 지역 출신이었기 때문에 서로 간의 건강한 사회적 유대를 증진시키기 위해서는 건물 형태 측면에서 우호적이고 개방적인 환경이 필요했다.

날씨 보호 요소는 주로 하루 종일 주변에 있는 공간을 즐길 수 있도록 작업이 이루어졌다. 한낮 이후의 햇볕은 서쪽으로 노출된 건물 표면을 가장 뜨겁게 한다. 이 문제를 해결하기 위해 캔틸레버 슬래브 위에 벽을 비뚤비뚤하게 쌓아 필요한 보호벽을 만들었다.

이 과정에서 건물의 형태 및 언어와의 통합이라는 아이디어가 생겨났다. 휴식을 위한 슬래브로부터 벽을 캔틸레버로 만들 수 있다면 어떨까? 마치 벽들이 정면부에서 튀어 나와 중앙 프레임 구조 위에서 균형을 잡는 것처럼 말이다. 이 플라잉 월은 시각적으로 매력적인 프레임을 만드는 발코니에 흥미로운 그림자를 드리운다. 주변 환경과 함께 묘사될 때 약간의 대조는 항상 유지되면서도 거주자들의 정체성을 형성하고 자부심과 소속감을 불러일으킬 정도로 압도적이거나 병합하지는 않는다.

Neighborhood Facility

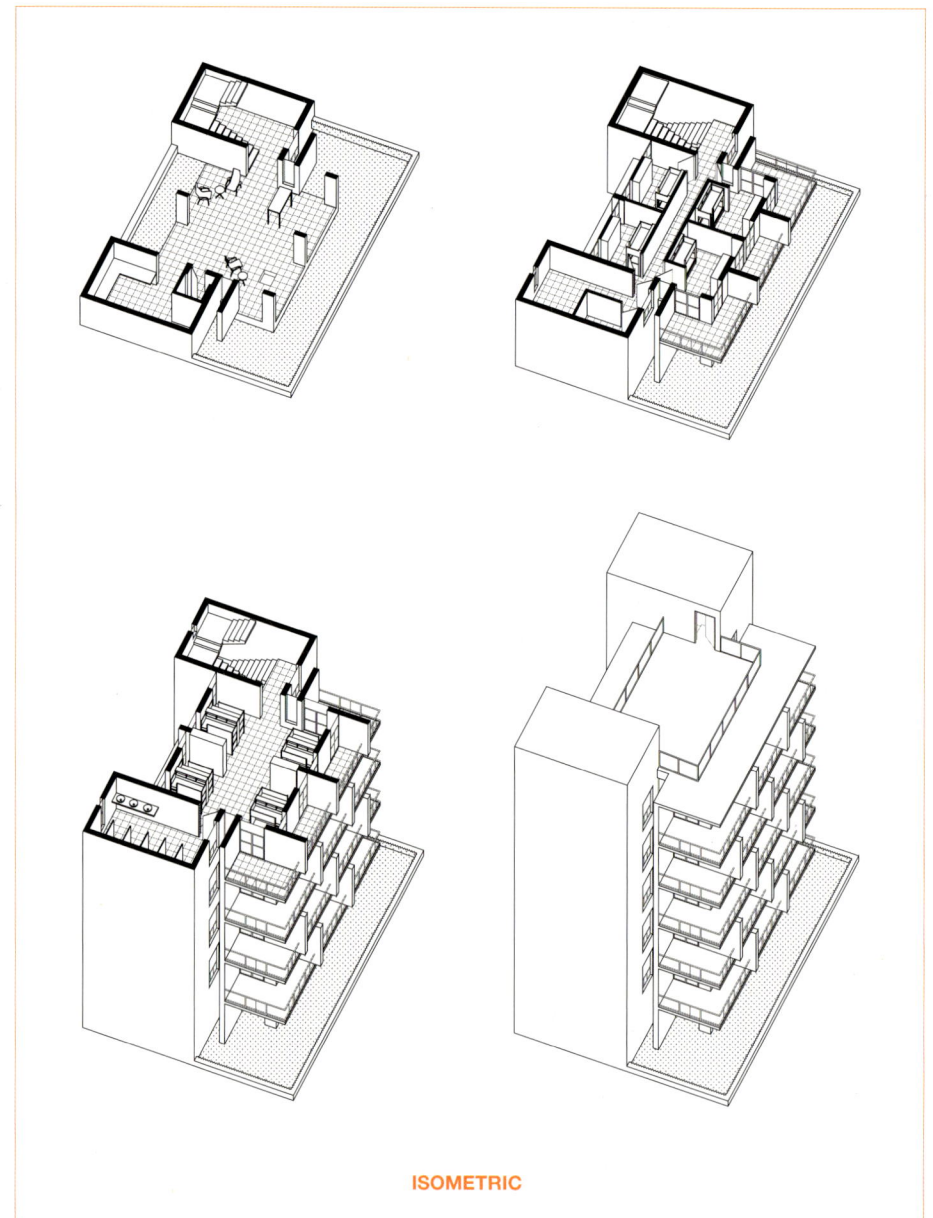

ISOMETRIC

ISOMETRIC VIEW-RENDERED   **FRONT VIEW-RENDERED**

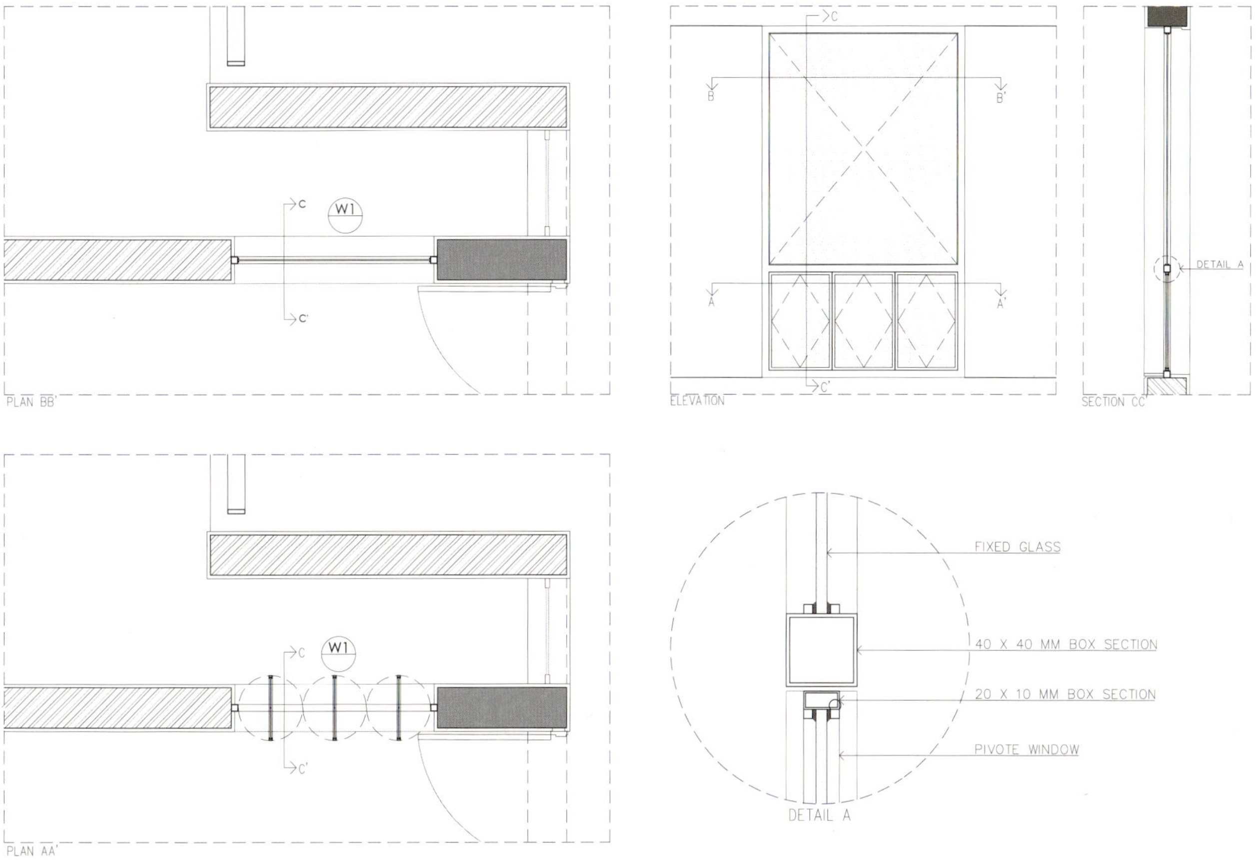

DETAIL

SECTION

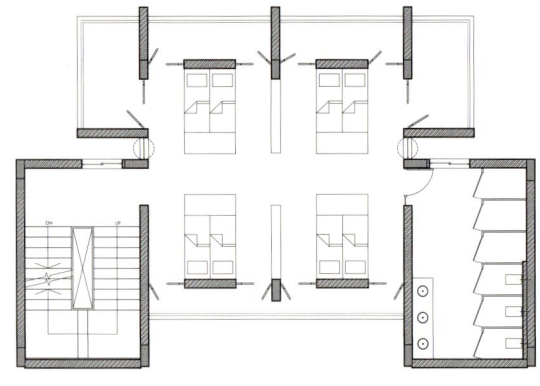

4TH FLOOR PLAN

TERRACE PLAN

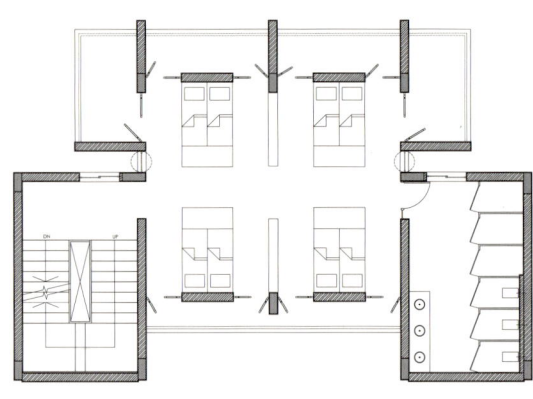

2ND FLOOR PLAN

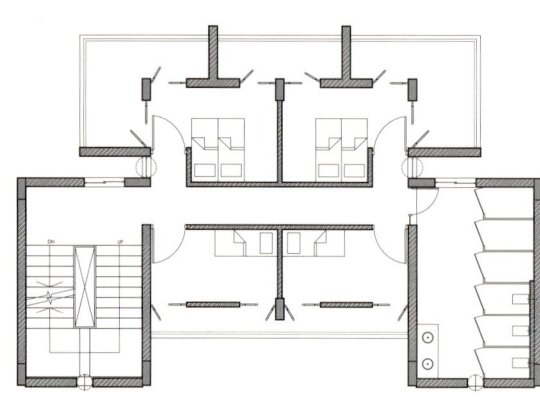

3RD FLOOR PLAN

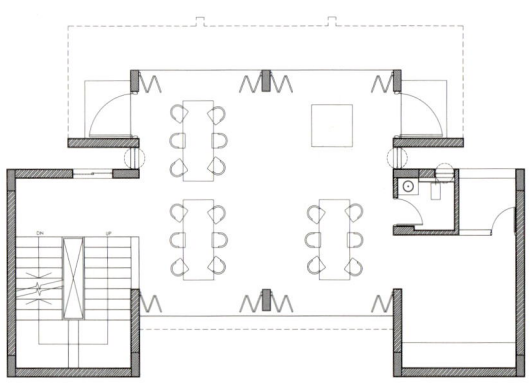

GROUND FLOOR PLAN

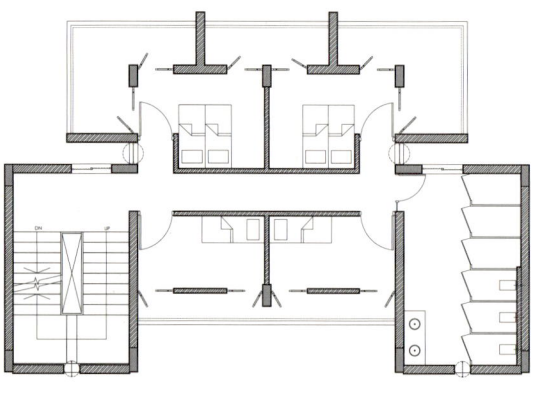

1ST FLOOR PLAN

Neighborhood Facility

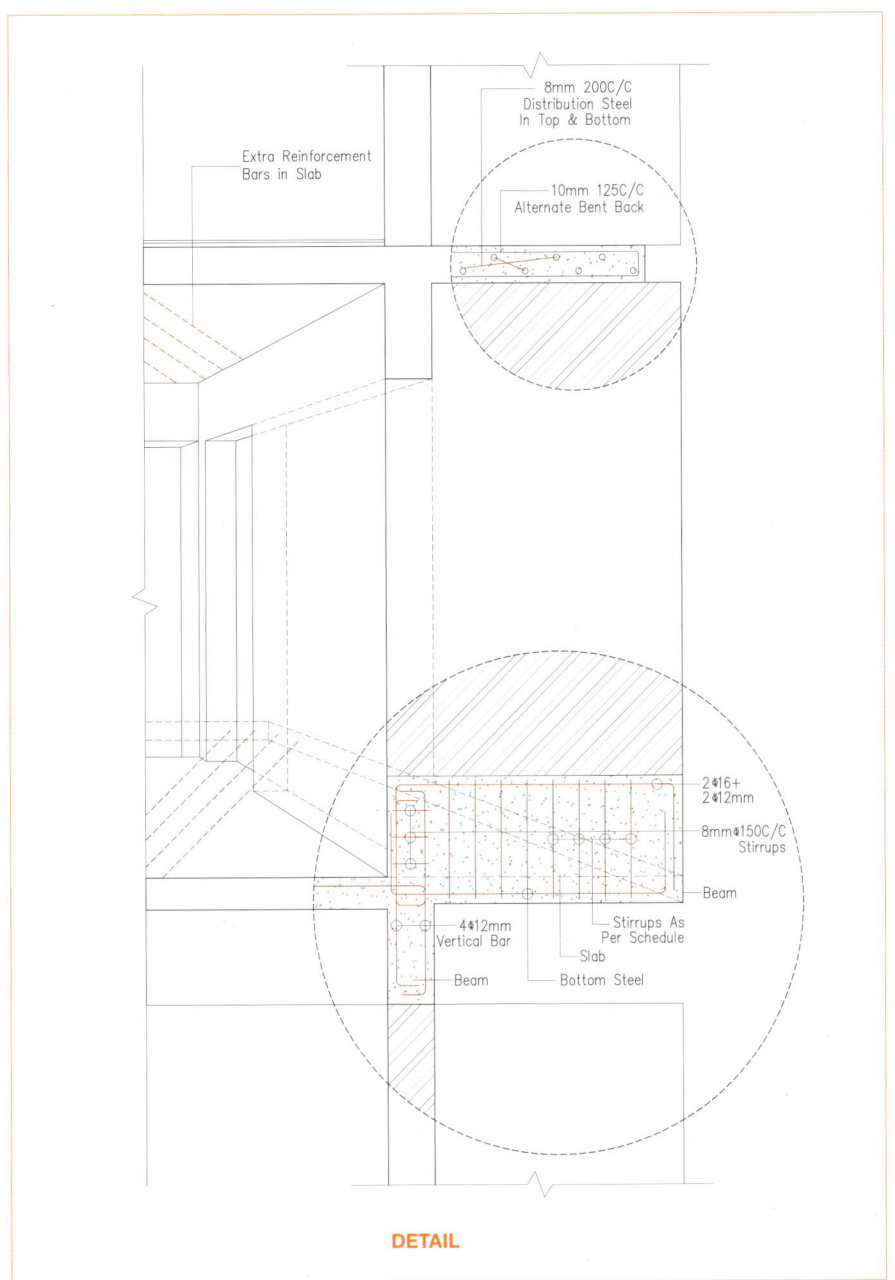

DETAIL

Komal Dhulia / Dhulia Architecture Design

Architect Komal Dhulia is the founding principal and the design director of the design-build firm that works across architecture, master planning, landscape, interiors and furniture. The firm is currently engaged in projects ranging from private residences, industries, institutions and commercial spaces.

'Dhulia Architecture. Design' was established in 2019 with a sincere effort toward building designs that belong to its time, place and identity. With a very personal touch, we grow and engage in the creation of context-appropriate designs.

The firm's work is focused on bringing unconventional ideas to reality, keeping in mind the ever-changing nature of the world. We customarily strive to achieve a vital balance in the spaces we deliver. The materials we employ and the details they form; speak of harmony.

>> https://studiodhulia.in/

Neighborhood Facility

Location
Hanoi, Vietnam
Use
Café
Site area
197m²
Built area
134m²
Total floor area
510m²
Main materials
Tile Viglacera Dong Anh (20,000 pieces), concrete, steel, glass

Project team
Doan Thanh Ha, Tran Ngoc Phuong, Luong Thi Ngoc Lan, Tran Van Duong, Nguyen Hai Hue, Ho Manh Cuong, Nguyen Van Thinh, Trinh Thi Thanh Huyen
Photographer
Le Minh Hoang

Ngói Space

타일 공간

H&P ARCHITECTS

Neighborhood Facility

the Tree + the Cave = a big Roof

SKETCH

Neighborhood Facility

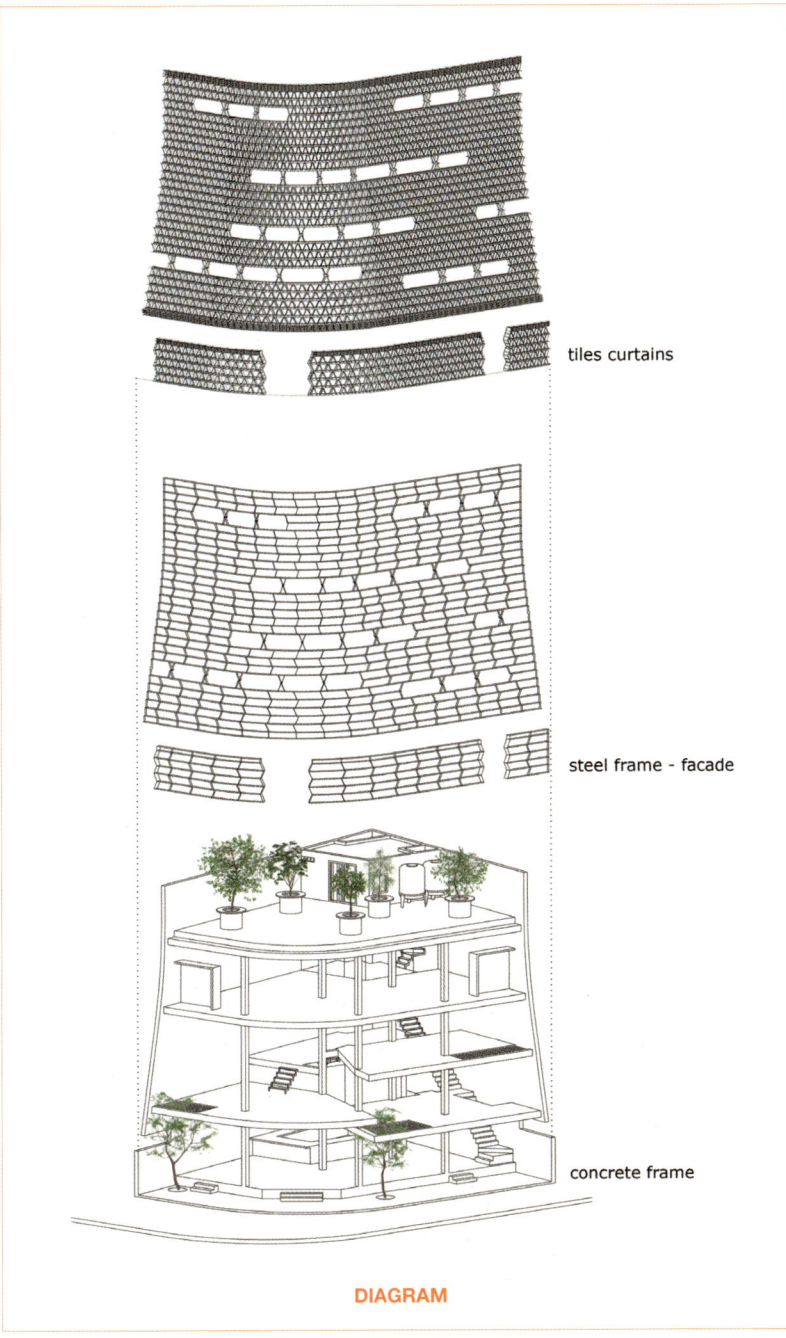

tiles curtains

steel frame - facade

concrete frame

DIAGRAM

Architecture can be, in a sense, analogous to a Tree (banyan tree, bodhi tree) with its branches and leaves that intertwine and spread out, forming various layers for use; It can also be metarphorically meant spatial levels offering miscellaneous chambers at different heights in a Cave, as regards prehistoric men being accommodated.

Ngói space is created from the perspective of merging these 2 primitive shelters (the Tree & the Cave), giving reminiscences of a big Roof such as the roof of a communal house or that of a Rong house – an open community space which has been existing for a long time.

In making a Roof, Tile presents a natural option since it is a material familiar with most Vietnamese people, yet used in an unusual way (as walls, curtains) in an extraordinary space to produce a special different effect (regarding appearance, sense of space and micro-climate quality). The Ngói space functions Cafés (lower space + on the rooftop), Seminar and Exhibition rooms (on the middle floor) with the aim of offering a fresh destination, an inspiring common Roof.

The on-going rapid urbanization and increasing population have been entailing a major demand for more areas for residence nationwide. Many former single-story tiled-roof houses have been demolished/removed and the tiles on those roofs are still considered construction waste, not to be reused. The Ngói space was created as an inspiring solution to reusing these memory-filled tiles. On a larger scale, it orientates users towards a sustainable tomorrow, from the perspective of reaching back to the past to recognize and rediscover the core and hidden values of the original space and use those values to create spaces of the future.

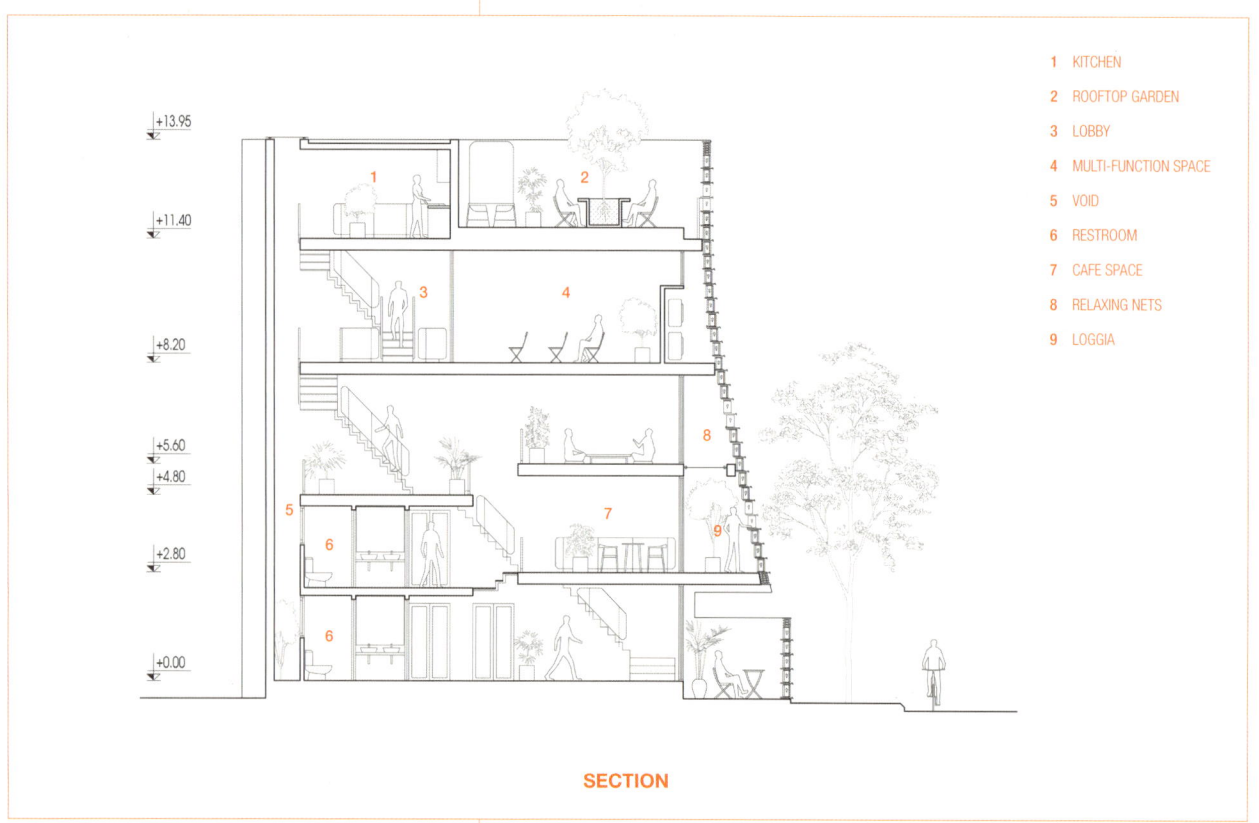

1	KITCHEN
2	ROOFTOP GARDEN
3	LOBBY
4	MULTI-FUNCTION SPACE
5	VOID
6	RESTROOM
7	CAFE SPACE
8	RELAXING NETS
9	LOGGIA

SECTION

Neighborhood Facility

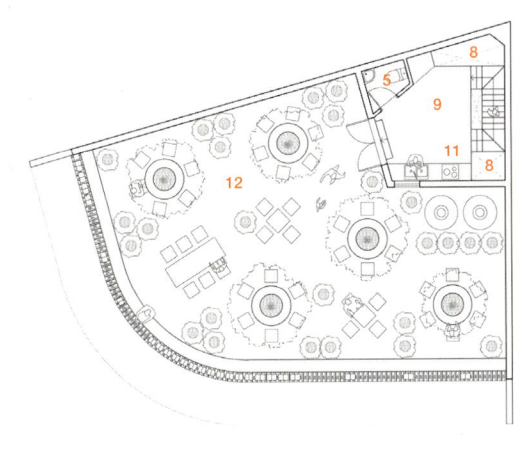

ROOFTOP GARDEN

1 ENTRANCE	7 RELAXING NETS
2 CAFE SPACE	8 VOID
3 DISPENSING COUNTER	9 LOBBY
4 STORE	10 MULTI-FUNCTION SPACE
5 RESTROOM	11 KITCHEN
6 LOGGIA	12 ROOFTOP GARDEN

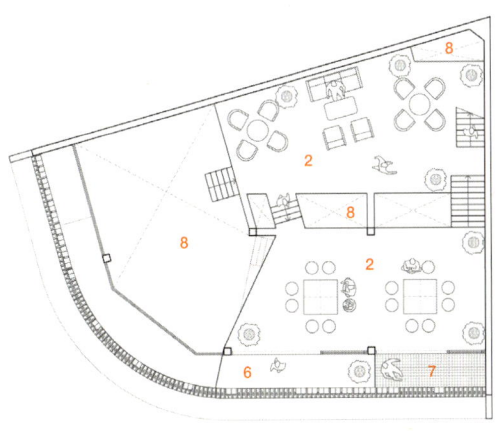

3RD FLOOR PLAN

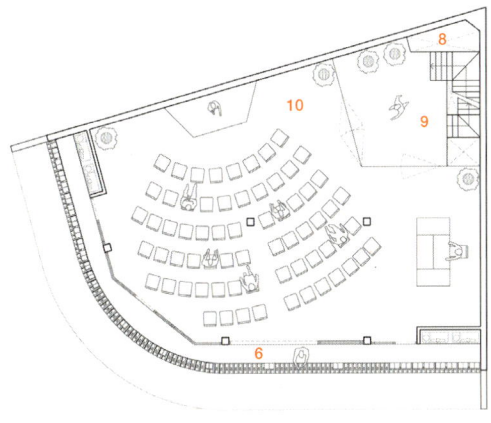

4TH FLOOR PLAN

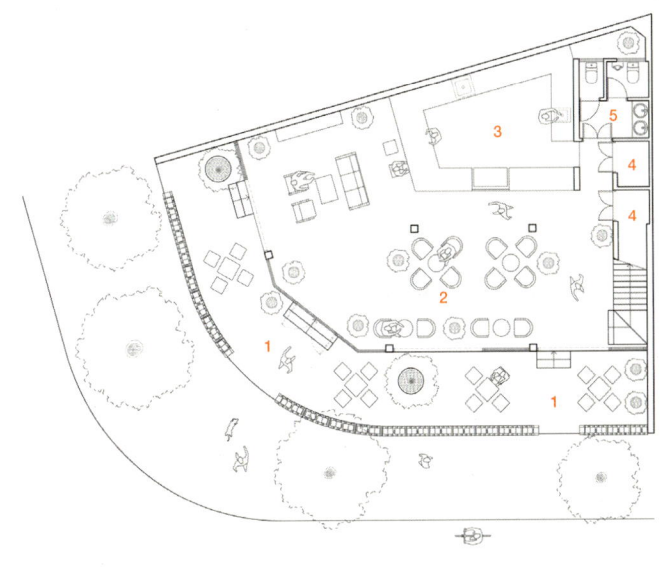

1ST FLOOR PLAN

2ND FLOOR PLAN

건축은 어떤 의미에서는 가지와 잎이 얽히고 설켜 다양한 층을 형성하는 나무(반얀트리, 보리수)와 유사 할 수 있다. 또한 선사시대 사람들이 거주하던 동굴에서 다양한 높이의 여러 가지 방을 제공하는 공간적 층위를 은유적으로 의미할 수도 있다.

Ngói space는 이 두 가지 원시적인 쉼터(나무와 동굴)를 융합하는 관점에서 지어졌으며, 주민회관 지붕이나 롱 하우스(Rong house) 지붕과 같이 오랫동안 존재해 온 개방형 커뮤니티 공간의 큰 지붕을 연상시킨다.

타일은 베트남 사람들에게 친숙한 재료이기 때문에 지붕 소재에 자연스러운 선택이지만, 특별한 공간에서 독특한 효과(외관, 공간감, 미세 기후 적합 품질)를 낼 때는 이색적인 방식(벽, 커튼 등)으로 사용된다. Ngói space는 카페(저층 공간 + 루프탑), 세미나실, 전시실(중간층)로 구성되어 있으며 신선한 느낌의 목적지로서 모두를 위해 영감을 주는 지붕이 되는 것을 목표로 한다.

계속되는 급속한 도시화와 인구 증가로 인해 전국적으로 주거 공간에 대한 수요는 늘어나고 있다. 이전의 많은 단층 기와지붕 주택은 철거 및 해체되었으며, 이들 기와는 여전히 재사용이 불가한 건설 폐기물로 여겨지고 있다. Ngói space는 이러한 추억이 담긴 타일의 재사용을 위한 영감을 주는 솔루션으로 창조되었다. 나아가 과거로 돌아가 원래 공간의 핵심 가치와 숨겨진 가치를 인식하고 재발견하며, 이러한 가치를 바탕으로 미래 공간을 창조한다는 관점에서 지속 가능한 미래를 지향한다.

Neighborhood Facility

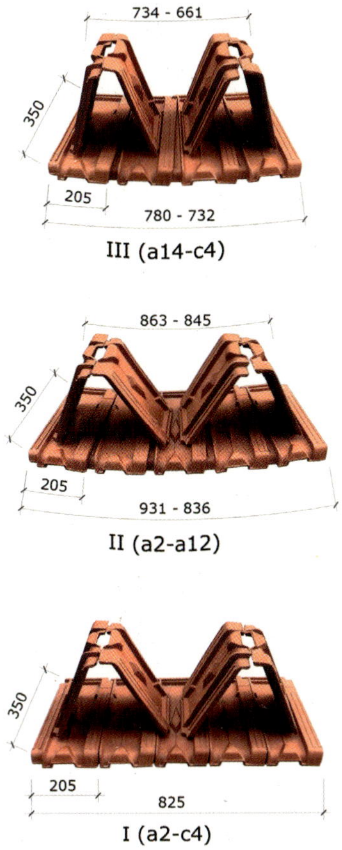

TYPICAL UNITS

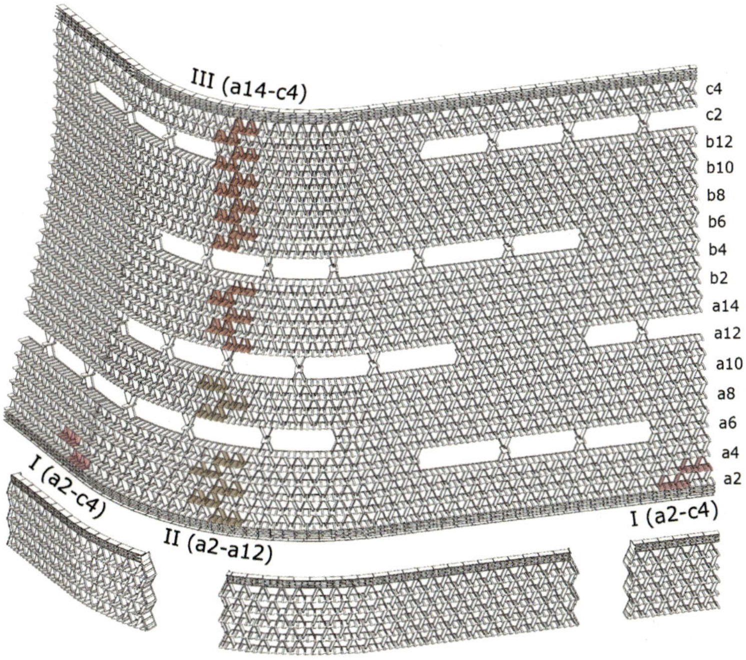

TILES CURTAINS DETAIL

267

Neighborhood Facility

Neighborhood Facility

Neighborhood Facility

Doan Thanh Ha / H&P Architects

Doan Thanh Ha graduated from Hanoi Architectural University in 2002, in 2007 he holds a masters degree in Hanoi Architectural University. He set up and have been operating H&P Architects since 2009.
His works have been receiving high appraisals in Vietnam and overseas and honourable prizes including UIA Turgut Cansever International Award, UIA Friendly and Inclusive Spaces Award, ARCASIA Awards for Architecture, National Architectural Award,...
>> www.hpa.vn

Location
Tokyo, Japan
Use
Shop
Site area
163.58m²
Built area
138.27m²
Total floor area
231.62m²
Floors
4F

Interior finish
Wall_ LGS base+Reinforced gypsum board +AEP coating on the joint cheesecloth, Floor_ Concrete finishing with a trowel+Dust-proof clear finish, Furniture_ Clear tempered glass+Cutting board+Old wood+Steel finishing with melamine resin coating
Project architect
KAMITOPEN Co., Ltd.

Partner company
Naganuma Architects Co., Ltd., EIGHT BRANDING DESIGN Co., Ltd.
Photographer
Keisuke Miyamoto

KAMA-ASA Shop

가마아사(釜浅) 상점

KAMITOPEN CO., LTD.

Neighborhood Facility

Kamaasa Shoten opened in 1909 when Minosuke Kumazawa founded "Kumazawa Foundry" in Kappabashi, Asakusa. Since then, the foundry has been confronting and dealing with chefs and a wide variety of tools for over 100 years.

This time, it is a store refurbishing plan for the 4th generation of a shop owner. This project started in 2011. At the beginning, we used deformed steel bars as materials resembling raw metal and produced fixtures that imitated the shape of an iron kettle. Since then, for 10 years, the store has been fulfilling its mission as a place of a spacial encounter point between customers and displayed cooking utensils. From that point, we have been proceeding with this reconstruction plan.

Concept "the rebar"

The rebar is a stick-shaped rolled steel material formed from irregular nubs called ribs and joints that are used as a building construction material.

As KAMA-ASA Shop's policy is set on belief that good tools are made for good reason, it delivers to the customers excellent skills of craftsmen, which make kitchen utensils. Therefore, the most appropriate for the shop's space was to show architectural skills of craftsmen. The rebar that is usually hidden in the wall became the main design. Handrails and furniture were designed with rebar. Although they are simple, they still require the use of craftsman skills.

The knives' shelves are made from railroad ties. They express the long history of knives that can be used for many years.

As designers, we hope that craftsmen skills will be introduced to many people.

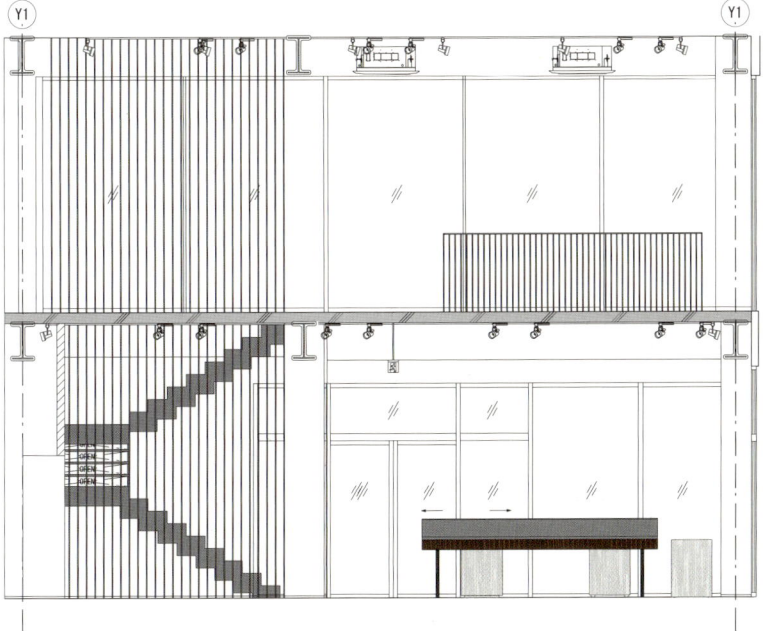

SECTION

가마아사(釜浅) 상점의 시초는 1909년 구마자와 미노스케가 아사쿠사 갓파바시에 설립한 "구마자와 주조소"이다. 이후 이 주조소는 100년이 넘도록 요리사들에게 필요한 다양한 도구들을 제조해 왔다. 이 프로젝트는 상점주의 4대손을 위한 매장 리뉴얼 프로젝트로 2011년에 시작되었다. 처음에는 변형된 철근을 자재로 사용해 철제 주전자의 모양을 본뜬 고정물을 만들었다. 이후 10년 동안 매장은 고객과 진열된 조리도구의 공간적 만남의 장소로서 그 사명을 다해 왔다. 그 시점부터 우리는 이 재건축 계획을 진행해 왔다.

"리바" 콘셉트
리바(rebar)는 막대 형태의 압연 강재로, 건물 건설재로 사용되는 조인트와 리브라고 하는 불규칙한 혹이 있다.

KAMA-ASA Shop의 정책은 좋은 공구는 쓸모 있게 만들어, 주방 용품을 만드는 장인의 뛰어난 기술을 고객에게 선사하는 것이라는 믿음에 바탕을 두고 있다.
따라서 작업장 공간에 가장 적합한 것은 장인의 건축 기술을 보여주는 것이었다. 일반적으로 벽 속에 숨어 있는 리바가 메인 디자인이 되었다.
핸드레일과 가구를 리바로 설계했다. 단순하지만 장인의 기술이 필요하다.
칼 선반은 철도 침목으로 만들었다. 이 선반은 여러 해 사용할 수 있는 칼의 오랜 역사를 표현한다.
작업에 참여한 건축가로서 장인의 기술이 여러 사람들에게 소개되기를 바란다.

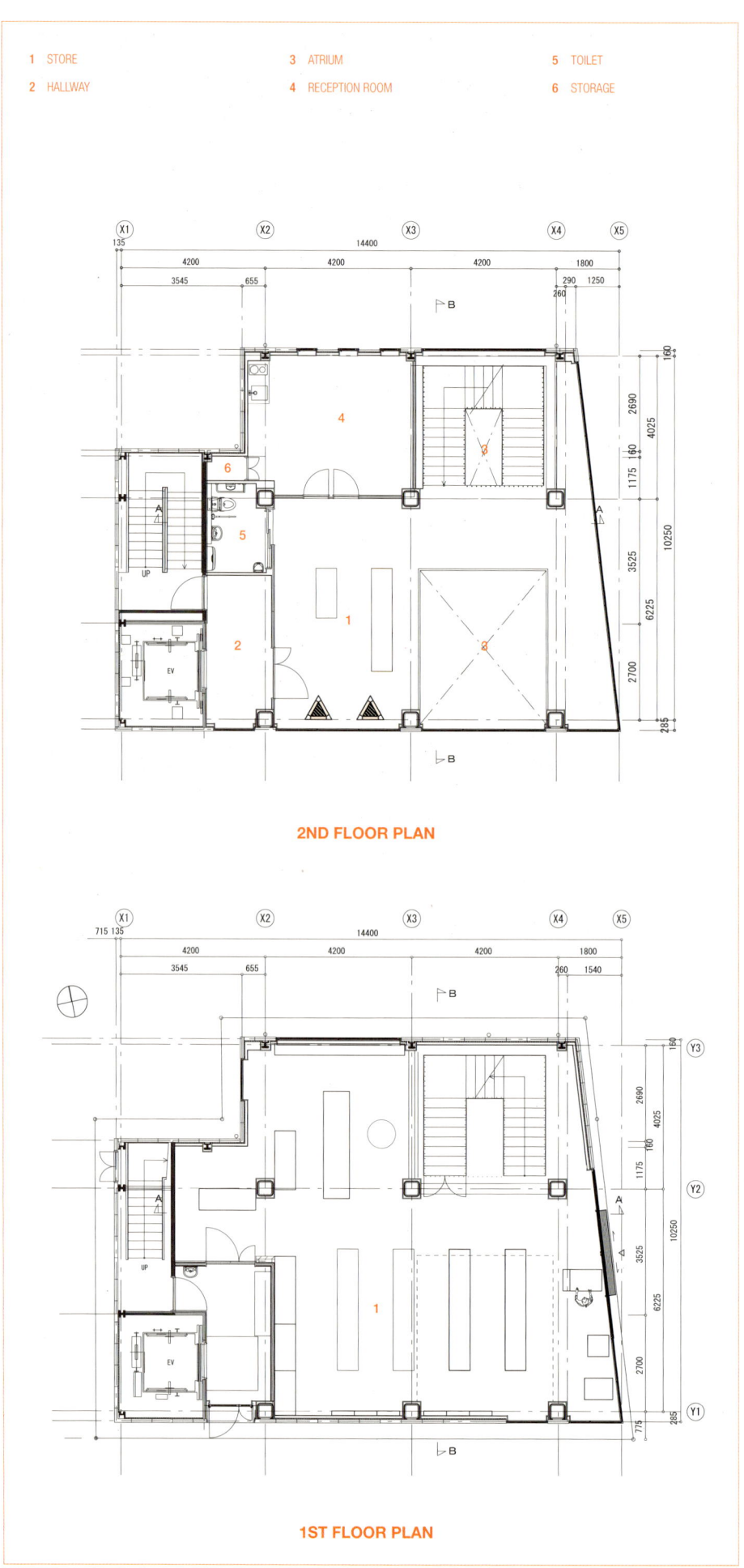

1 STORE
2 HALLWAY
3 ATRIUM
4 RECEPTION ROOM
5 TOILET
6 STORAGE

2ND FLOOR PLAN

1ST FLOOR PLAN

Neighborhood Facility

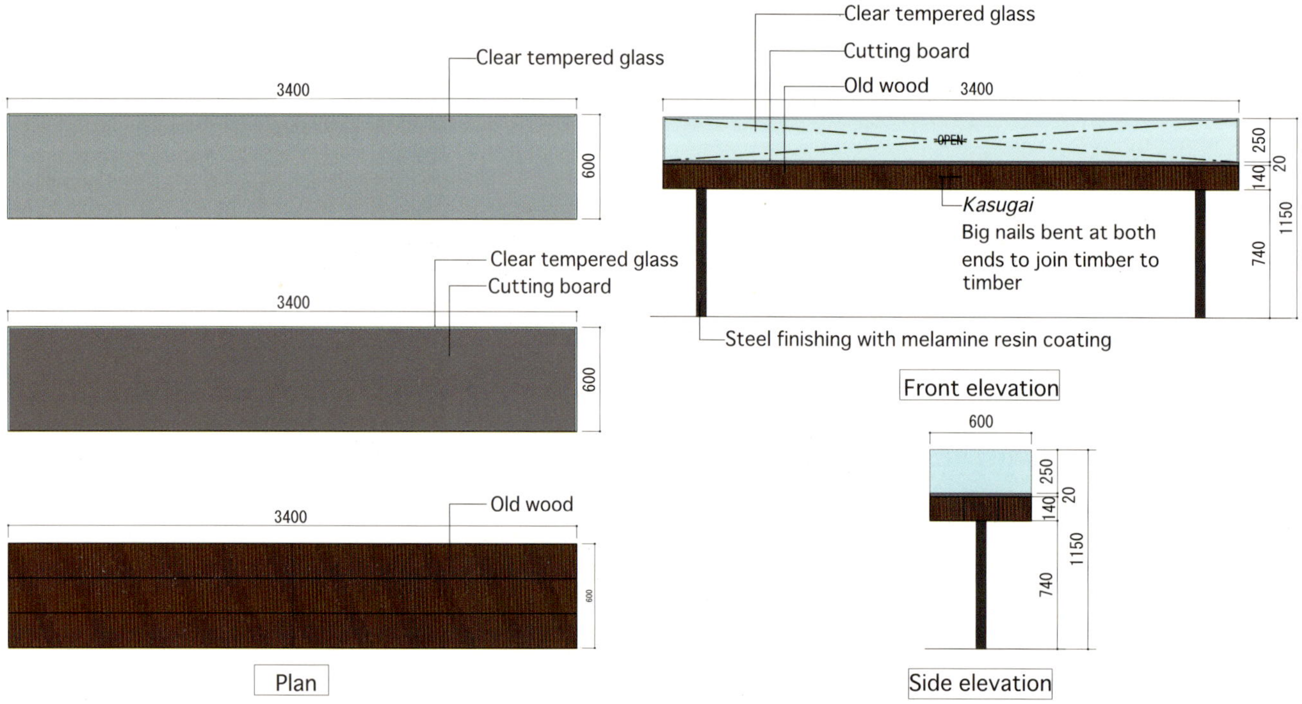

Plan

- 3400 × 600 — Clear tempered glass
- 3400 × 600 — Clear tempered glass / Cutting board
- 3400 × 600 — Old wood

Front elevation

- Clear tempered glass
- Cutting board
- Old wood
- 3400
- 250 / 140 / 20 / 1150 / 740
- *Kasugai* — Big nails bent at both ends to join timber to timber
- Steel finishing with melamine resin coating

Side elevation

- 600
- 250 / 140 / 20 / 1150 / 740

KITCHEN KNIFE STAND

Neighborhood Facility

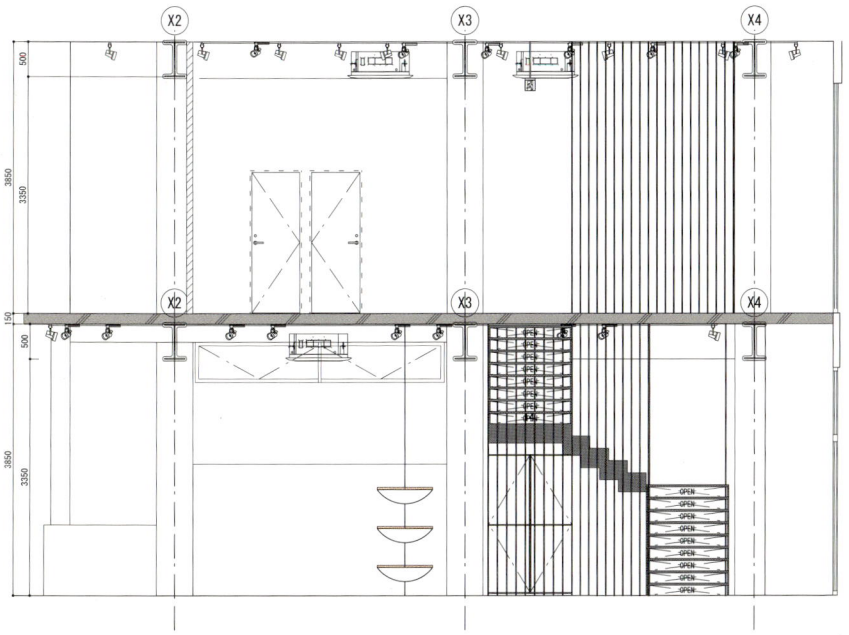

SECTION

KAMI TO PEN

KAMITOPEN Co., Ltd.

"From thought to reality to connection"
Clients have their own ideas to express to the public. The ideas can't be expressed without being converted into "words". They also can't be passed on without being transformed into "forms".

We are the designers who create the "words" and the "forms".
By putting "thoughts" into "words", which morph into "forms", a diverse range of people will be able to read and understand the ideas of the client.
This is the "design" that we aim for.
>> kamitopen.com/en/

Location
Seocho-gu, Seoul, Republic of Korea
Use
Commercial
Site area
315.60m²
Built area
142.67m²
Total floor area
835.16m²
Floors
B1, 5F

Exterior finish
Steel
Interior finish
Paint
Project architect
Lee Keun Sik
Construction
Fine Construction & Development
Photographer
Gu Ui jin

SUNU BLDG.

선우빌딩

엘케이에스에이 건축사사무소
LKSA

BEFORE

프로젝트는 양재천을 마주보고 있는 대지의 90년대 완공된 건물을 대수선하여 새로운 생명을 부여하는 것이다. 대수선의 주된 항목은 전체 입면의 형태 및 재료의 재구성이다. 기존 건물이 양재천을 끼고 있는 도로변으로부터 후퇴되어 있었다. 따라서 인접 건물들보다 인지성이 낮았기 때문에, 기존 건물 입면의 개구부 비율을 높임으로써, 낮과 밤에 좀더 부각이 되는 방향으로 계획하고자 하였다.

또한 전체 건물의 누수를 비롯한 하자 요소를 개선하는 것이다. 기존 입면이 드라이비트로 구성되어 있었고, 곳곳에 균열이 존재하고, 재료의 노후화가 진행되어 외부 우수의 침투에 취약한 상태였다. 최상층 두겁 후레싱도 노후화되어 우수침투의 하자를 가속화 시키는 요인으로 작용하고 있었다. 따라서 기존 드라이비트를 전량 철거하고, 창호방수와 골조 균열부등을 개선하고, 온전한 새로운 입면을 구성하고자 하였다. 커튼월 디자인은 기존 건물의 수직/수평의 비율을 파악하여 정분할 하여, 온전한 입방체의 정제미를 구현하려는 방향으로 진행하였다.

30년 이상 자신의 도시적 기능을 묵묵히 수행해온 건물이, 다시 건강하고, 고결한 모습으로, 변화된 도시와 사람들의 삶의 한 가운데에서 긍정적인 역할을 차분하게 해나가기를 바라본다.

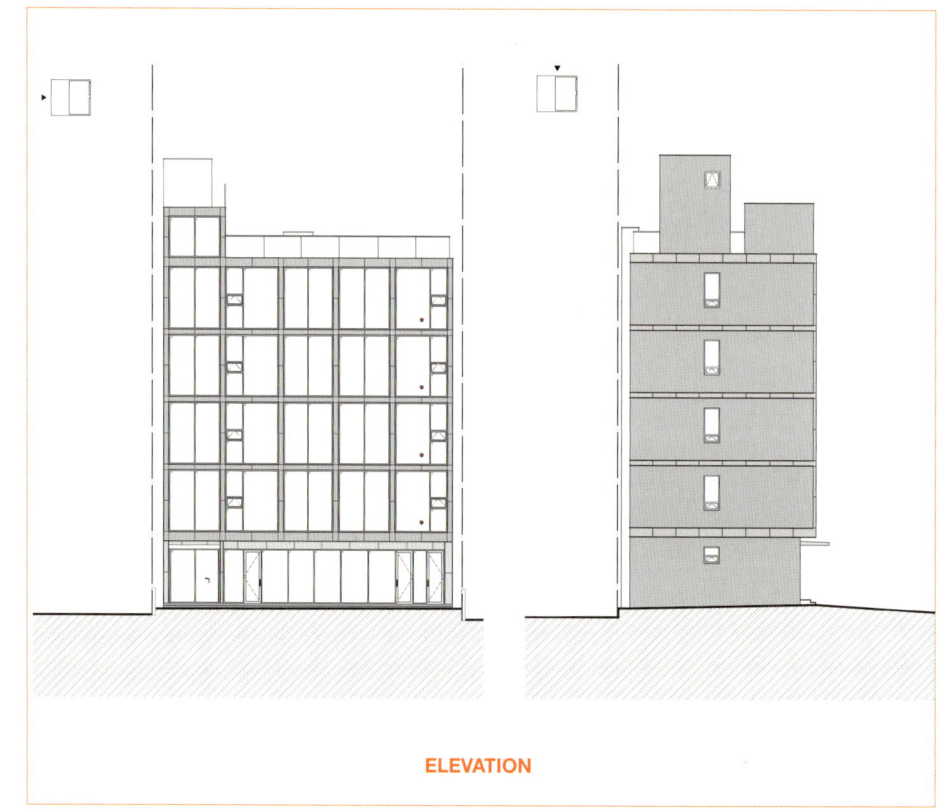

ELEVATION

Neighborhood Facility

DISMANTLING/ FACADE DEMOLITION (해체 / 입면철거)

STRUCTURAL REINFORCING / FACADE MOCK-UP TEST (철골보강 / 입면 목업테스트)

FACADE CONSTRUCTION (입면설치)

The project was to bring new vitality to a building completed in the 1990s in a location facing Yangjaecheon Stream. The rebuilding of the whole façade's shape and materials was the main focus of the refurbishment. The original building was set back from the edge of the roadside running along Yangjaecheon. Since it was less recognizable than the neighboring buildings, it was designed to be more noticeable during the day and at night by increasing openings on the original façade.

The project was also to fix defects, including leaks all over the building. The existing façade was made up of Dryvit, and cracks were found all over it as the progressive aging of the materials had made it vulnerable to rainwater infiltration from the outside. The edge flashing on the roof top was also deteriorating, causing an acceleration of the defects caused by rainwater penetration. Therefore, the plan was to remove the entire existing Dryvit, improve the waterproofing on the windows and doors, and address the frame cracks to create a whole new façade.

A curtain wall design was developed to bring out the exquisite beauty of perfect cubes by calculating the vertical and horizontal ratios of the original building and splitting them evenly.

We hope that the building, which has served its urban function for more than 30 years, will play a positive role in the midst of the changes in the city and the lives of its inhabitants with its new healthy and noble look.

1 RESTAURANT
2 STAIR
3 TOILET
4 WATER TANK
5 SEPTIC TANK
6 OFFICE
7 MANUFACTURING STORE

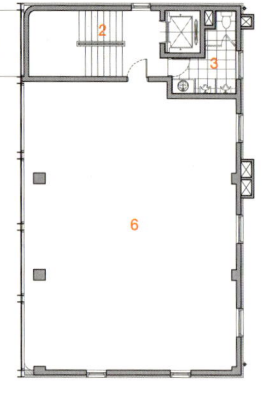

2ND FLOOR PLAN

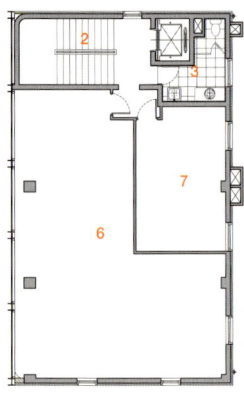

5TH FLOOR PLAN

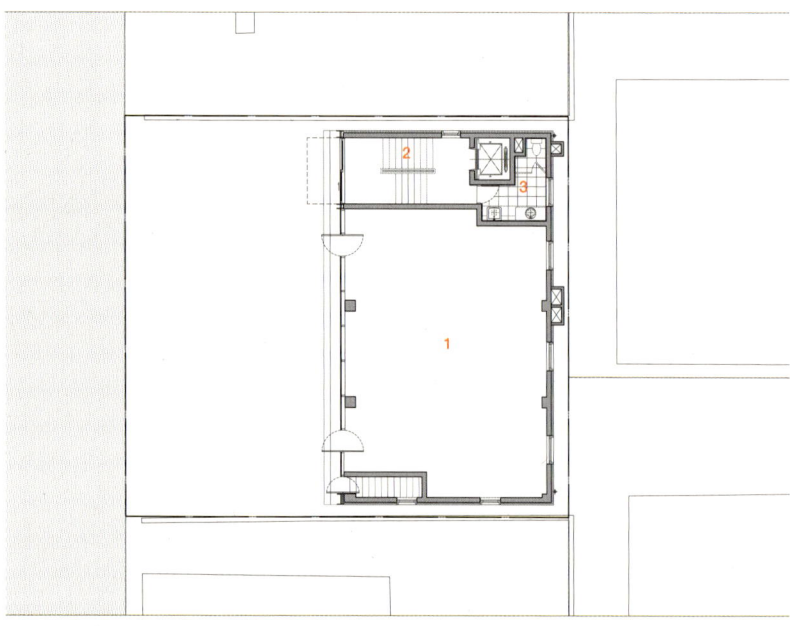

1ST FLOOR PLAN

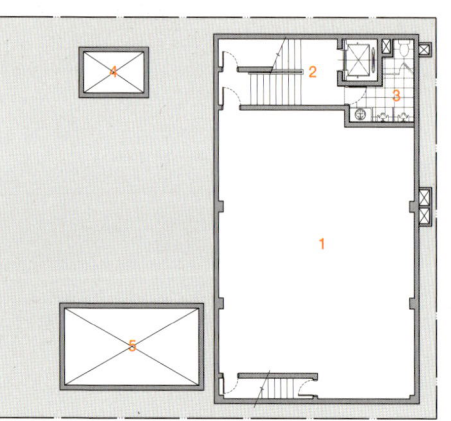

B1 FLOOR PLAN

Neighborhood Facility

이근식 / LKSA 건축사사무소
Lee Keunsik / Lee Keun Sik Architects

LKSA 건축사사무소는 '건축가의 본질적인 의무와 책임은 건축주를 비롯한 그 공간을 향유하는 사람들의 행복을 위한 노력에 있다'라는 신념 하에 2012년부터 매 순간 건축을 향한 깊은 애정과 장인정신 그리고 소명 의식을 갖고 설계한다. 건축뿐만 아니라 인테리어, 가구, 조경, 사업 컨설팅까지 건축에 관여하는 모든 요소의 전문성을 가지고 있다. 이 모든 요소가 건축가의 일관된 사고에서 연속성을 가질 때 비로소 삶을 위한 그릇이 현실화된다고 믿는다.

이근식 건축가는 한양대학교를 건축학과를 졸업하고 삼우종합건축사사무소에서 실무를 쌓았으며, 2012년부터 LKSA 건축사사무소를 운영해오고 있다. 2020년에 '대한건축사협회 신진건축사상'을 수상했다.

\>> www.lksa.kr

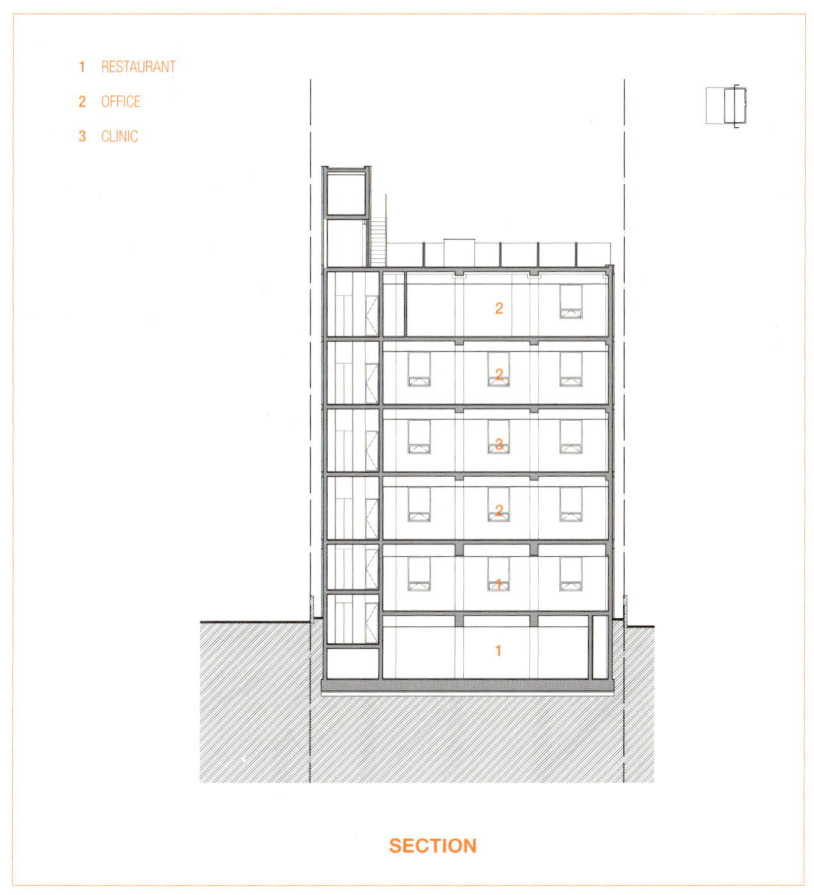

1 RESTAURANT
2 OFFICE
3 CLINIC

SECTION

Neighborhood Facility

Location
Mapo-gu, Seoul, Republic of Korea
Use
Commercial & office
Site area
277.9m²
Built area
155.28m²
Total floor area
725.82m²
Floors
B1, 5F

Exterior finish
Long brick tile, curtain wall
Interior finish
Exposed concrete, concrete polishing
Project architect
Park Nam-kyu
Construction
Gadam Construction Co.,Ltd.
Photographer
Kim Chang-mook

W 134 (INOSYS Headquarters Office)

더블유 일삼사 (이노시스 사옥)

(주)오엠엠건축사사무소
OMM ARCHITECTS

홍익대학교 앞 와우산로는 과거 미술학원들이 들어서 실기를 준비하는 미술입시생들로 인산인해를 이루던 길이었다. 홍익대학교 미술대학이 2013년 실기시험을 폐지하면서 하나둘 빠져나가기 시작한 가로는 사람의 보행이 현저히 줄어들고 지역은 활기를 잃어가고 있다. 건축주는 광고회사로 70%는 사옥으로 사용하고 30%는 임대목적으로 사용할 것이란 요구를 하였고, 따라서 상층부는 사무공간으로 저층부는 상업공간으로 사용될 수 있도록 고려했다.

본 대지는 정방형 형태로, 북측으로 전면 20m 도로와 동측면 6m 도로에 접해 있으며 와우산 방향으로 3m의 고저차를 가지고 있다. 지형을 이용해 지하 1층이 실질적인 지상 1층의 기능을 하고 주차 때문에 줄어든 지상 1층에 약 50%의 내부 보이드를 두어 지하 1층의 공간을 복층공간으로 활용할 수 있도록 하였다. 확장된 지하공간을 통해 도시가로에 개방감을 부여하고, 더불어 지역가로를 활성화 시킬 수 있기를 기대하였다.

전체 조형은 주변스케일을 고려해 매스를 분절시켜 스케일을 낮춤으로써 주변 보행자들에게 친밀도를 높일 수 있도록 고려했다. 사무공간은 전면도로인 북측은 창을 최소화하여, 업무의 집중도를 높일 수 있게 했으며, 남측으로 창을 두어 채광을 유입시켜 밝은 업무환경이 되도록 하였다. 특히, 5층 공간은 광고회사의 특성을 살려 다양한 이벤트가 가능한 공간으로 계획했으며, 조형적으로 유리블록을 삽입하여 홍대앞 어울마당로에서 접근하는 보행자들을 향한 비주얼 타깃이 되도록 함으로써 상징성을 부여하였다. 옥탑층은 도로측 가벽과 엘리베이터를 옥탑층까지 연장시켜 접근성을 높임으로써 평상시에는 직원들의 휴게공간으로 활용되고, 회사의 다양한 외부 이벤트도 수용할 수 있도록 계획하였다.

DIAGRAM

Neighborhood Facility

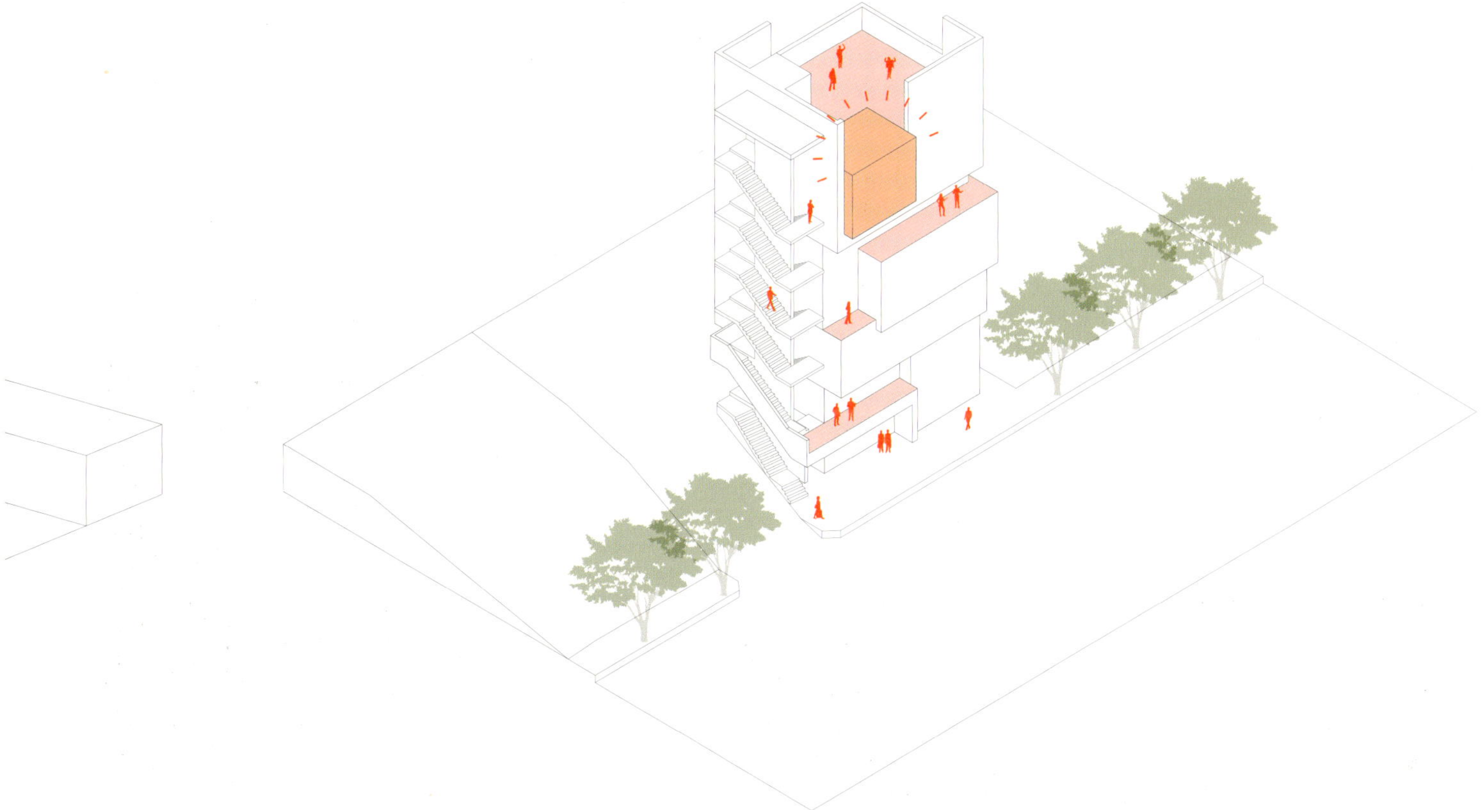

Wausan-ro in front of Hongik University used to be a street crowded with art academies and art students preparing for practical skill exams. As Hongik University's Fine Arts Department abolished the practical exam in 2013 and academies began to leave the street, the people traffic substantially declined, and the neighborhood was losing its vitality. The client, an advertising firm, requested that 70% of the building space be used for its offices and 30% for rental purposes, so it was considered that the upper levels would be used as office space and the lower levels as commercial space.

The land is square-shaped, is bordered by a twenty-meter-wide road on the north side and a six-meter-wide road on the east side, and the elevation is three meters higher in the direction of Wausan Mountain. The first basement level may be utilized as a multi-story space since it was designed to serve as the first floor above ground utilizing its topography, and around 50% of the internal void was given to the first floor, whose area was diminished due to the parking lot. It was expected that the expanded underground

Neighborhood Facility

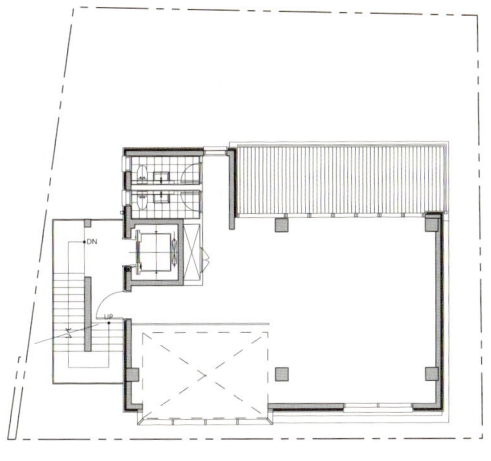
5TH FLOOR PLAN

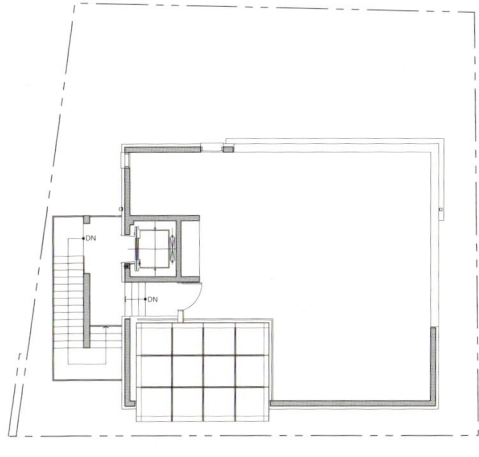
ROOF FLOOR PLAN

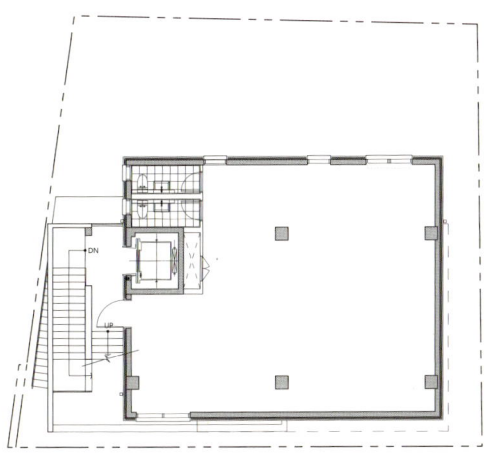
2ND FLOOR PLAN

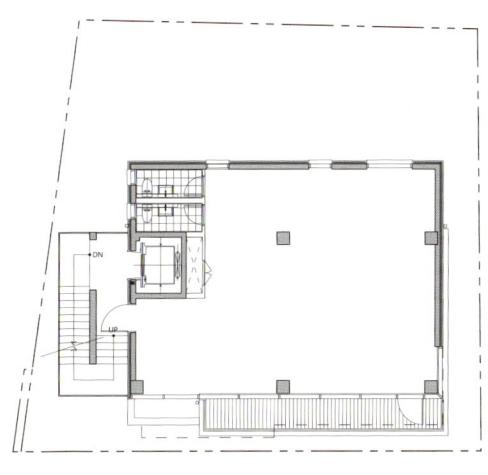
4TH FLOOR PLAN

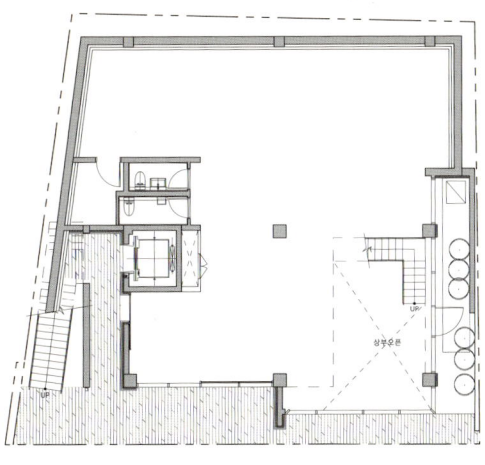
B1 FLOOR PLAN

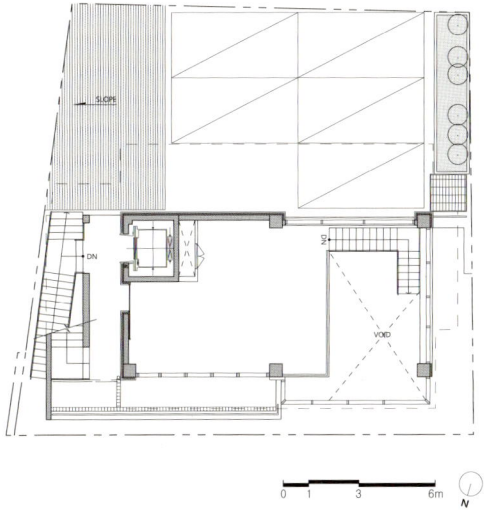
1ST FLOOR PLAN

space would open up the city streets and energize the neighborhood streets.

For the overall structure, the idea was to reduce the scale by dividing up the mass in consideration of the surrounding scales and thereby provide an inviting feeling to visitors and pedestrians. For the office space, windows were minimized on the north side, facing the frontal road, to block distractions from work, while windows were placed on the south side to bring light in and create a bright work environment. The fifth floor was specially designed to be a venue for a variety of activities, considering the distinctive nature of an advertising firm. In addition, glass blocks were added in a formative way to offer symbolism as a visual draw for pedestrians coming from Eoulmadang Street in front of Hongik University. The space on the rooftop was planned to serve as a resting area for employees during regular times and to accommodate a variety of outside events for the company as the roadside walls and elevator extending to the rooftop increase accessibility.

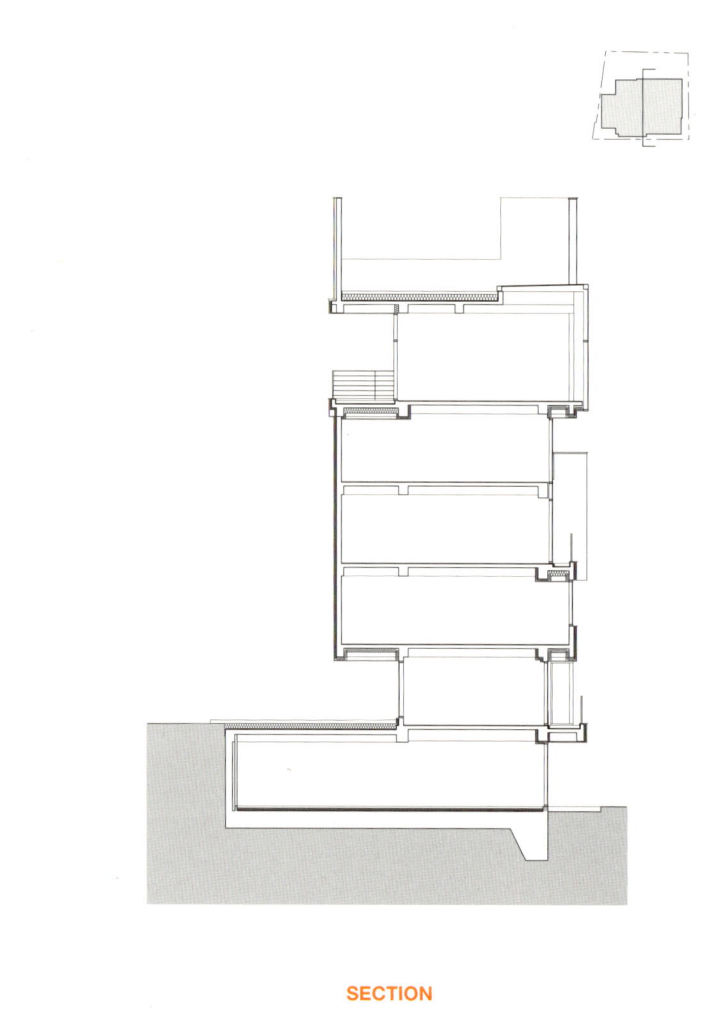

SECTION

김원영, 박남규, 이종희 / (주)오엠엠건축사사무소
Kim Wonyoung, Park Namkyu, Lee Jonghee /
omm architects

오랜 시간 공간건축에서 근무해온 3명의 소장이 함께 건축설계사무소 topos를 창립으며, 2023년 omm으로 사명을 변경했다.
오엠엠건축사사무소 대표 건축가들은 각 13년~20년의 시간동안 공간건축에서 주택, 근린생활시설부터 대규모 공공턴키, 현상설계까지 충분히 다양한 경험을 축적하였다. 이들은 이 기간을 새로운 건축을 하기 위한 수련의 과정이었다고 생각한다.
건축은 이미 거장의 시대를 넘어 협력과 통섭을 기반으로 새롭게 성장하고 있다. 따라서 omm은 한 사람의 독자적 아이디어에 따르기보다는 다양한 사람, 지식, 문화, 기술 분야와의 협력과 수평적 조직을 지향한다. omm은 보다 새롭고 진보적인 도시의 공공성과 건축주의 의도에 기여하는 것을 목표로 하고 있다.

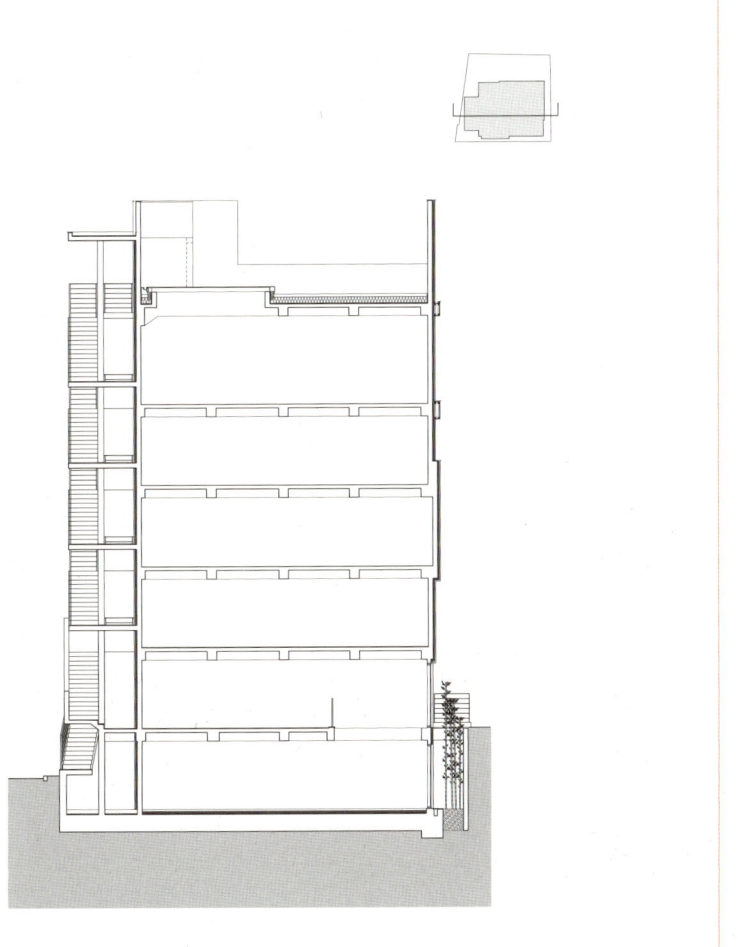

SECTION

Location
Tehran, Iran
Use
Office
Site area
340m²
Built area
180m²
Total floor area
2,035m²
Floors
B2F, 6F

Exterior finish
G.F.R.C Panel, Curtain Wall, Metal Louver, Metal Plates, Glass Handrail
Interior finish
Knauf panel, Rock Ceramic.
Project architect
Shahab Alidoost, Sona Eftekharazam
Photographer
Mohammad Hassan Ettefagh

Design Team
Amir Niknafs, Hamideh Raofzadeh, Sahand Mohaddes, Hamed Bakhtiari, Ali Taghibeygi, Venus Entezami, Ilia Salek, Ida Ehsani, Mina Mehrdad, Mohammad Ali Izadi, Ehsan Danandeh

VIRA II

비라 II

ALIDOOST AND PARTNERS

Neighborhood Facility

"VIRA II" office building is located in Yusefabad neighborhood at the central part of Tehran. This neighborhood is practically the host of both administrative and residential land uses and is assumed among the highly important and valuable zones of Tehran city. This building has been constructed with a plot area of 340m2 and has a total gross area of 2035m2 including a ground floor, six floors with administrative use and useful area of 180m2 in each floor, and two basement floors. The parking lots are located on the basement and ground floors.

The main problem in designing this project is its adjacence to old buildings on the western side, on the one hand, and neighboring a modern and high-rise building on the eastern side, on the other hand, and how to deal with this issue. Coordination with the two adjacencies differs both in terms of Function and in terms of form and scale. The western neighbors of the project are apartments mostly with residential use and 3 and 4-storey height, which are assumed as small-scale structures in Tehran metropolis and fully obey the municipal criteria in terms of form. In other words, these buildings have been constructed in full compliance with the dominant typology of the urban texture of Tehran. On the eastern side, this building is adjacent to a modern and high-rise building with administrative-commercial use, which is evidently different from the western neighbors in terms of scale, function, and form.

In dealing with the abovementioned problem, the primary challenge of the design team was the existence of an imposed bevel on the western side and avoidance from the form-related obedience to the typology resulting from the compliance with the criteria. Therefore, in the presented concept, the bevel not only has been utilized as a repetitive urban motif but also has gained a higher functionality through multiplication in volume and recess towards east in the central part of the building facade and serves as a tool for transforming the 2D face into a 3D active volume. The application of this approach and dividing the building into three overlapping volumes results in a better juxtaposition with the western neighbors in terms of scale. Furthermore, by using transparent surfaces in a major part of the main building facade and also by using repetitive balconies with a potential of locating green spaces, it has been attempted to establish a better coordination with the eastern neighbors.

As for the materials used in the building facade, it should be noted that considering the diversity of the materials in the neighboring buildings, the color functionality of the building has been reduced by using neutral and greyish colors in the project walls. The rigid parts of the southern face of the building is made of GFRC panels and metal sheets with dark gray coating color. The vertical elements and shaders contribute to the improvement of the climatic performance (controlling the southern and southwestern sunlight).
At the end, it must be noted that "VIRA II" attempts to have positive interactions with its surrounding urban sight, while maintaining its independent identity as a building, and avoid being indifferent to the urban landscape.

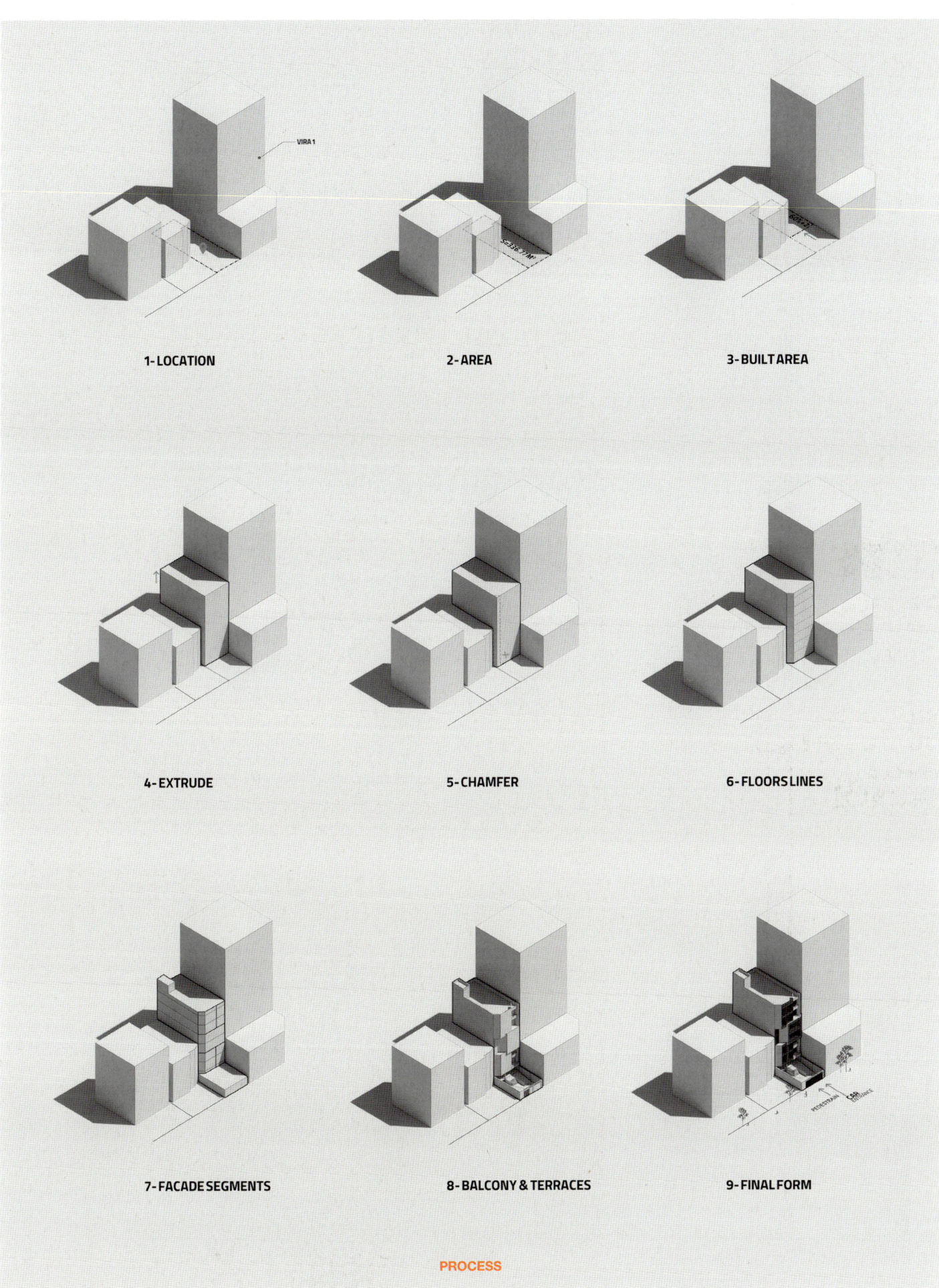

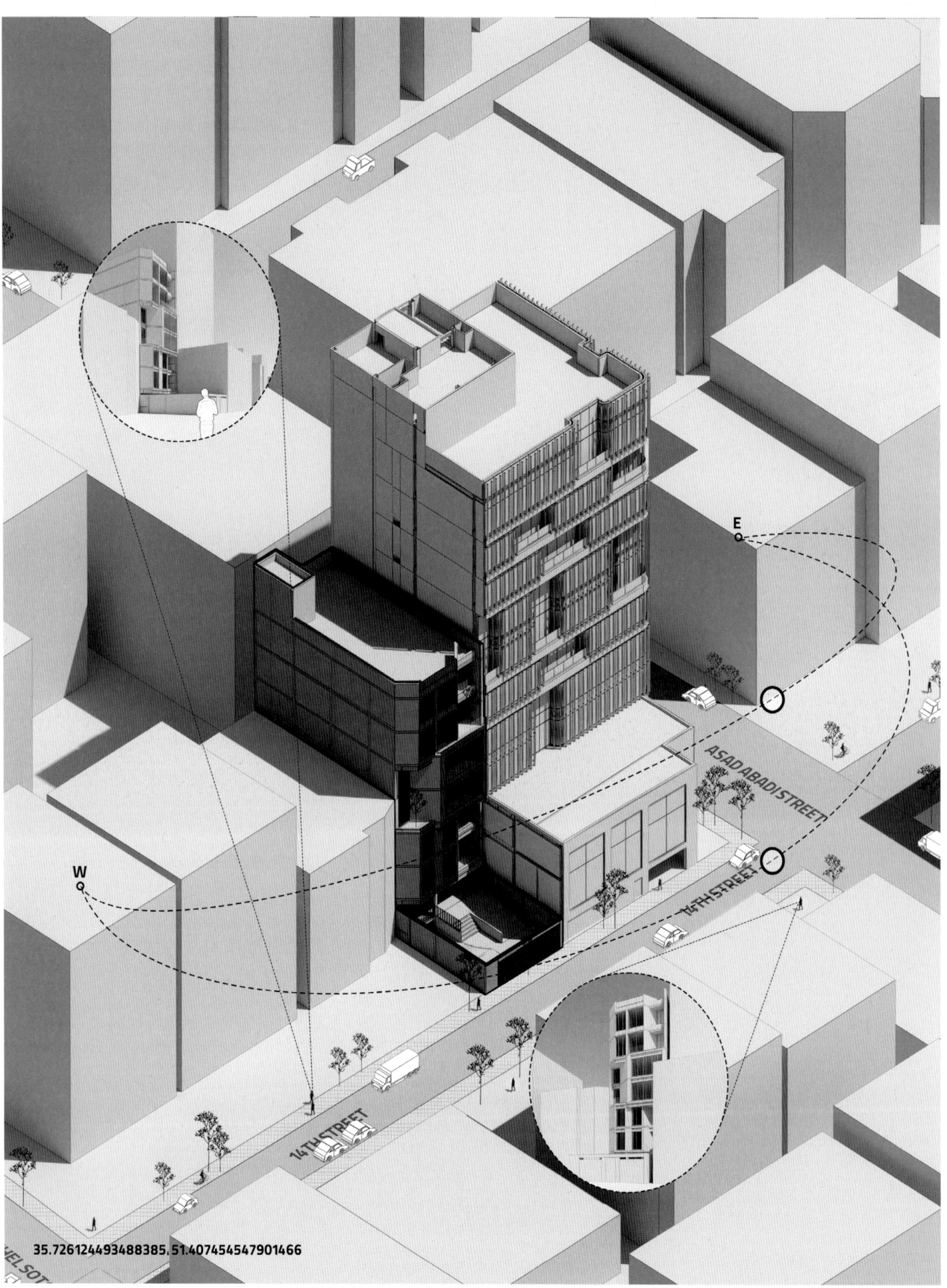

Neighborhood Facility

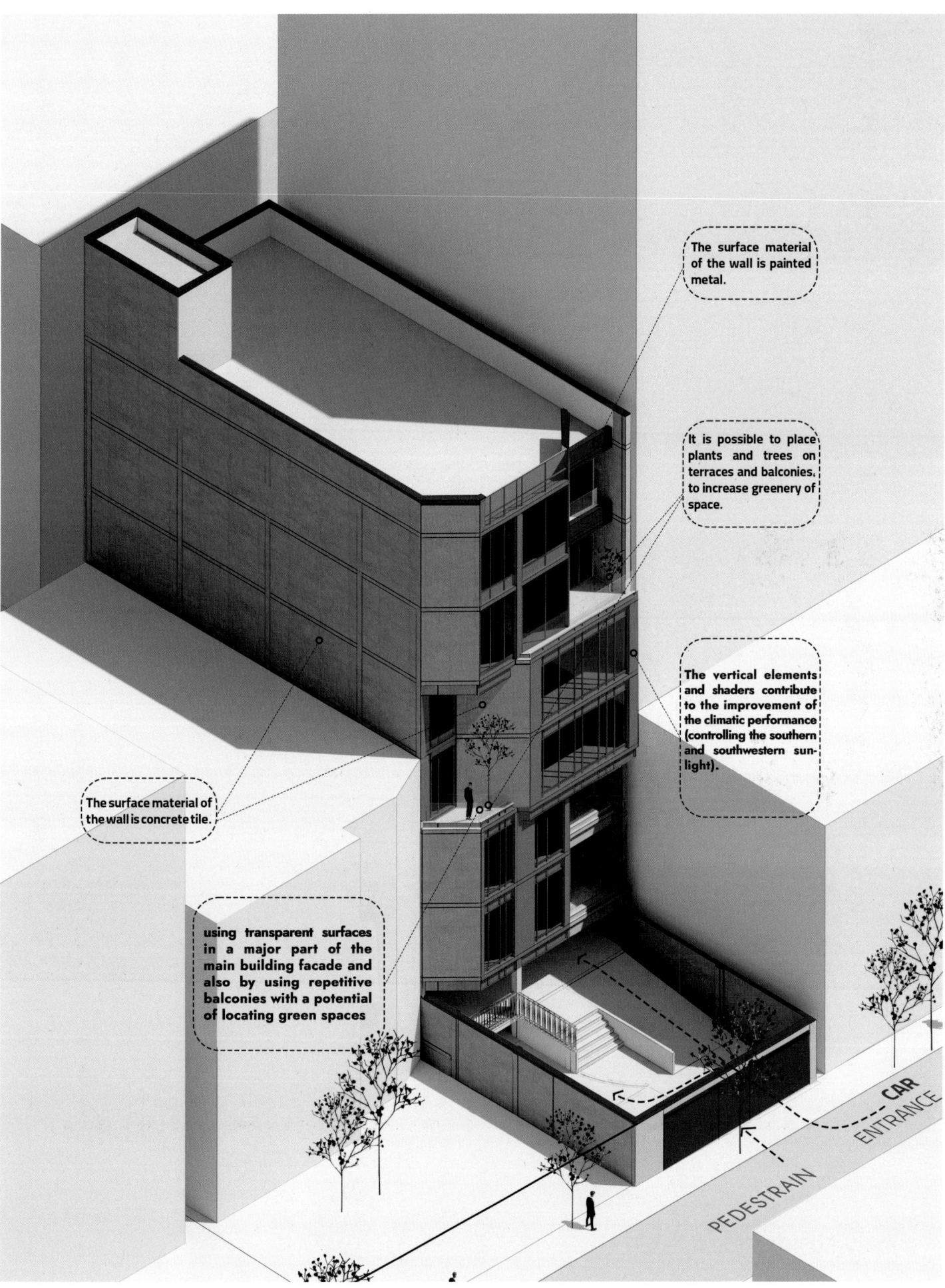

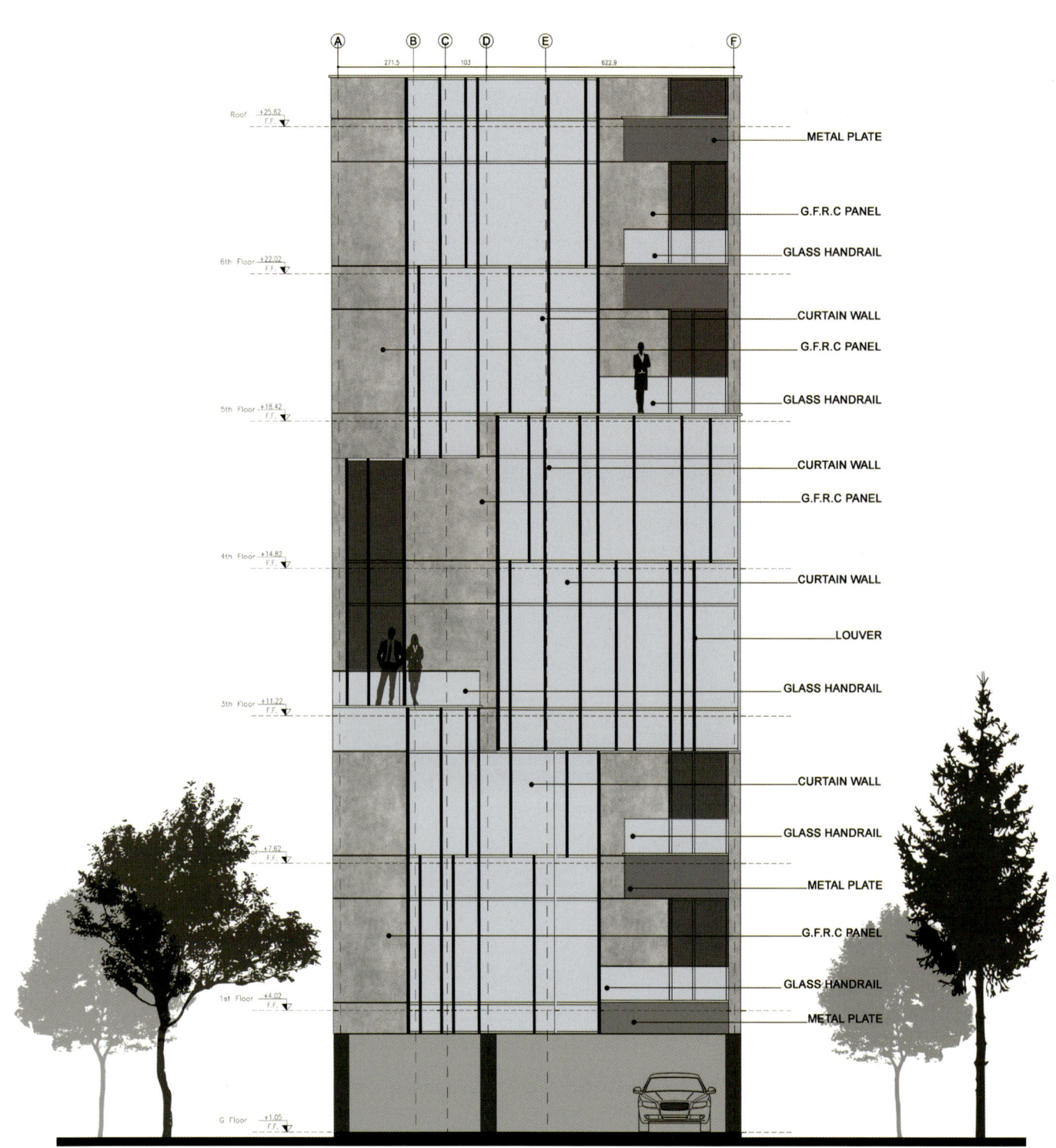

SOUTH ELEVATION DETAIL

Neighborhood Facility

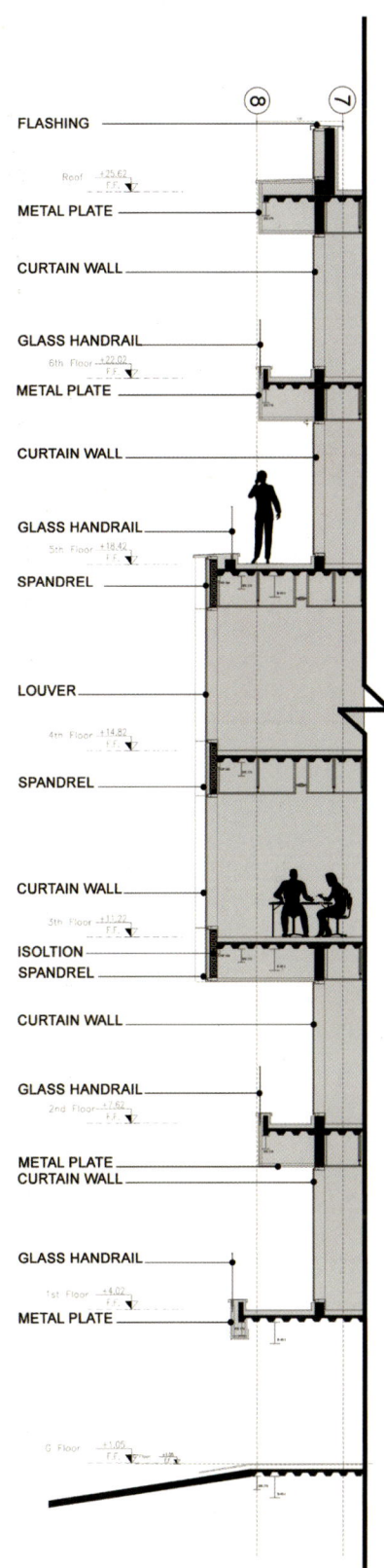

WALL SECTION DETAIL

"VIRA II" 오피스 빌딩은 테헤란 중심부인 유세프 아바드(Yusef Abad) 지역에 위치해 있다. 이 지역은 실질적인 행정 및 주거 지역으로, 테헤란에서 매우 중요하고 가치 있는 지역 중 하나이다. 340m² 부지에 지어진 이 건물은 총 연면적 2,035m²로, 층 구성은 1층, 층마다 180m² 면적의 관리/공용 공간을 갖춘 6개층, 지하 2개층으로 되어 있다. 주차장은 지하와 지상층에 있다.

이 프로젝트 설계에서 가장 큰 문제점은 건물이 서쪽의 오래된 건물들과 인접해 있는 동시에 동쪽의 현대식 고층 건물들과도 인접해 있어 이 부분을 어떻게 처리하느냐였다. 기능, 형태, 규모 면에서 차이가 있는 이 두 인접 지역들과의 조화가 필요했다. 이 프로젝트의 서쪽 지역은 주로 3~4층 높이의 주거용 아파트로 테헤란에서는 소규모 건물에 속하며 형태 면에서는 시 기준을 철저하게 준수한다. 즉, 이들 건물은 테헤란의 도시적인 느낌을 형성하는 지배적인 유형을 100% 준수하며 지어졌다. 이 건물의 동쪽은 행정-상업 용도의 현대식 고층 건물들이 인접해 있어 규모, 기능, 형태 면에서 서쪽과는 확연히 다르다.

이 같은 문제를 해결하기 위한 설계팀의 주요 과제는 서쪽으로 나 있는 경사면과 기준 준수에 따른 유형에 대한 형태적 순응을 피하는 것이었다. 그래서 제안된 콘셉트는 경사면을 반복적인 도시 모티브로 활용하는 동시에 건물 정면 중앙부에서 동쪽으로 움푹 들어가게 함으로써 체적을 배가시켜 기능성을 높이고 2D 면을 3D 활성 체적으로 변화시키는 도구로서의 역할을 하도록 하는 것이었다. 이 접근방식을 토대로 건물을 세 겹의 체적으로 분할하면 규모 면에서 서쪽 인접 건물과의 병렬 배치를 개선할 수 있다. 또한 본관 정면의 주요 부분에 투명한 표면을 사용하고, 그린 공간을 배치할 수 있는 반복형 발코니를 사용하여 동쪽 인접지와 조화를 이루고자 하였다.

건물 정면에 사용한 재료는 인접 건물의 재료 다양성을 고려하여 벽에는 중성적인 회색

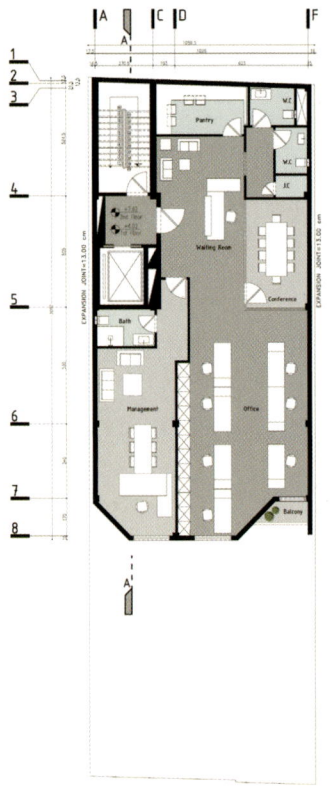

1ST AND 2ND FLOOR PLAN

3RD FLOOR PLAN

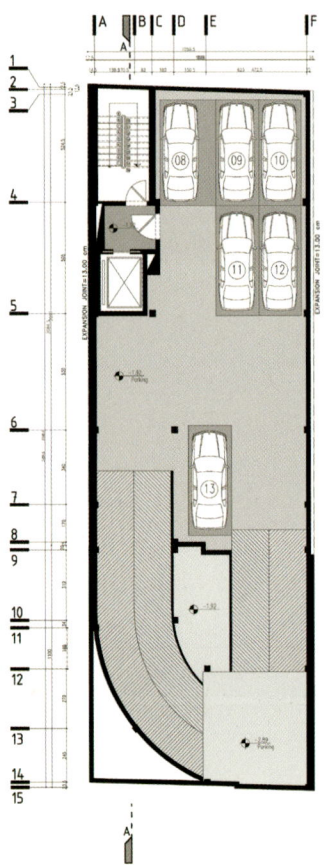

B1 FLOOR PLAN

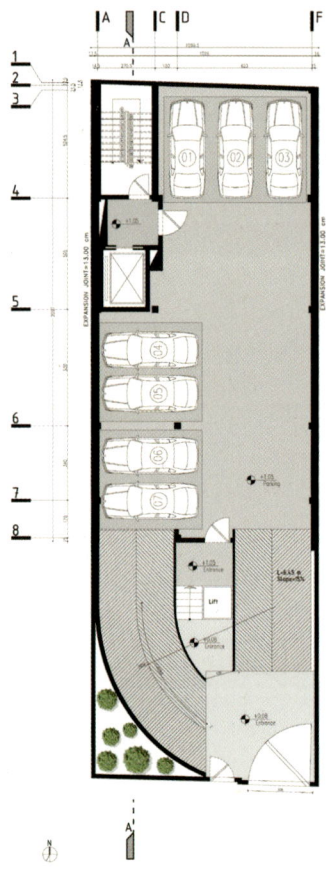

GROUND FLOOR PLAN

Neighborhood Facility

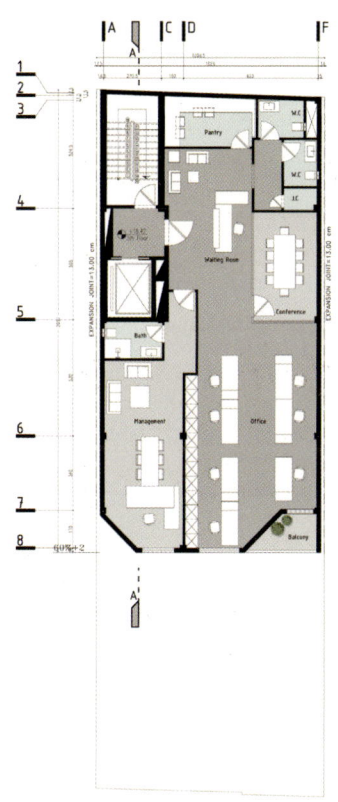

5TH FLOOR PLAN

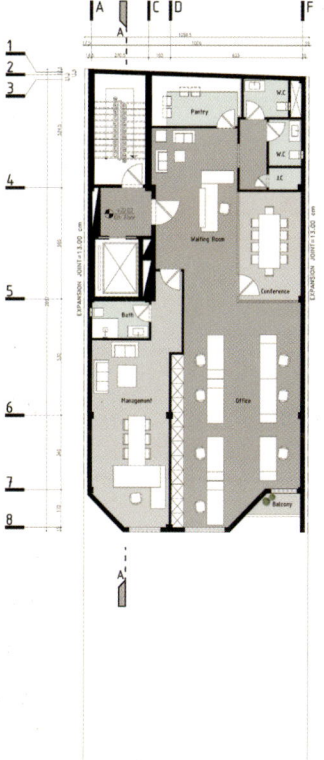

6TH FLOOR PLAN

계열의 색상을 사용함으로써 건물의 색상 기능성을 줄였다. 건물 남쪽 면의 단단한 부분은 GFRC 패널과 어두운 회색 코팅 색상의 금속 시트로 구성했다. 수직 요소와 셰이더는 기후 성능 향상에 기여한다(남쪽과 남서쪽 햇빛을 제어).

마지막으로 "VIRA II"는 건물로서 독립적인 정체성을 유지하는 동시에 주변 도시 경관과의 적극적인 상호작용을 시도하고 도시 경관을 최대한 존중하려 노력했다는 점에 주목해야 한다.

Shahab Alidoost + Sona Eftekharazam / Alidoost and Partners

Founded and directed by Shahab Alidoost and Sona Eftekharazam, Alidoost and Partners is a multidisciplinary office that focuses on architecture and design, from large scale planning to furniture. Rich with multiple expertises, our office is fuelled by talented designers and experienced architects that jointly develop projects from early sketches to on-site supervision. All of which, regardless of scale, outlines an approach that is affirmatively social in its outcome, enthusiastic in its ambition and professional in its process. At the core of our architecture is the ability to take a fresh look at design issues through experienced eyes. Our approach aims at turning intense research and analysis of practical and theoretical matters into the driving forces of design. By continuously developing rigorous methods of analysis and execution, Alidoost & Partners is able to combine innovative thinking and efficient production.

We work with corporate, government and private clients in numerous countries to realize major civic, hotel, residential, office, commercial and educational developments. We carefully limit the commissions we take on to help ensure a high degree of professional attention and overall project quality.

Alidoost &Partners envisions itself as a proactive partner for its client, rather than a consultant. The office has a wide portfolio of international work and the attitude of involving external consultants to improve the design intelligence of a given project team. The use of complementing teams ensures that a project will never suffer from being neither too conventional nor too naive.

>> www.alidoost-partners.com

Location
Bangkok, Thailand
Use/Program
Bar & Restaurant
Site area
185m²
Built area
185m²
Total floor area
385m²
Floors
Ground floor & mezzanine

Exterior finish
Aluminium composit, translucent roman tile
Interior finish
Laminate, wallpaper, polycarbonate, bronze mirror, stainless steel
Project architect
PAD Space Artisan
Contractor
S36
Photographer
Panoramic Studio

PENNANT Thonglor

레스토랑 페넌트 통로

PAD SPACE ARTISAN

Escape the bustle of city life for a moment and come aboard for a sublime dining experience.
Pennant, the new restaurant in Thonglor is founded by a group of F & B industry entrepreneurs who have gained international experience to develop the first flagship store for the Thonglor culinary scene, creating experiences customers won't soon forget.

Concept Design

The location of the existing building which locates at the beginning of Thonglor road facing the main Sukhumvit road is where our inspiration for the curves came from as it resembles the sail that is blown by the wind. This enlightened the creation of the translucent curves that bring dynamic and welcome patrons from Sukhumvit road. The first-floor facade is designed to contrast the second floor with PAD to elevate the scene to

EXTERIOR MATERIAL (FAÇADE)
A. Translucent Roman Tile : SCG
B. Alumimium Composit : SIAMBOND
C. Alumimium Composit : SIAMBOND
D. White Calacatta Marble : Stone Gallery
E. Terrazzo floor : by contractor

Neighborhood Facility

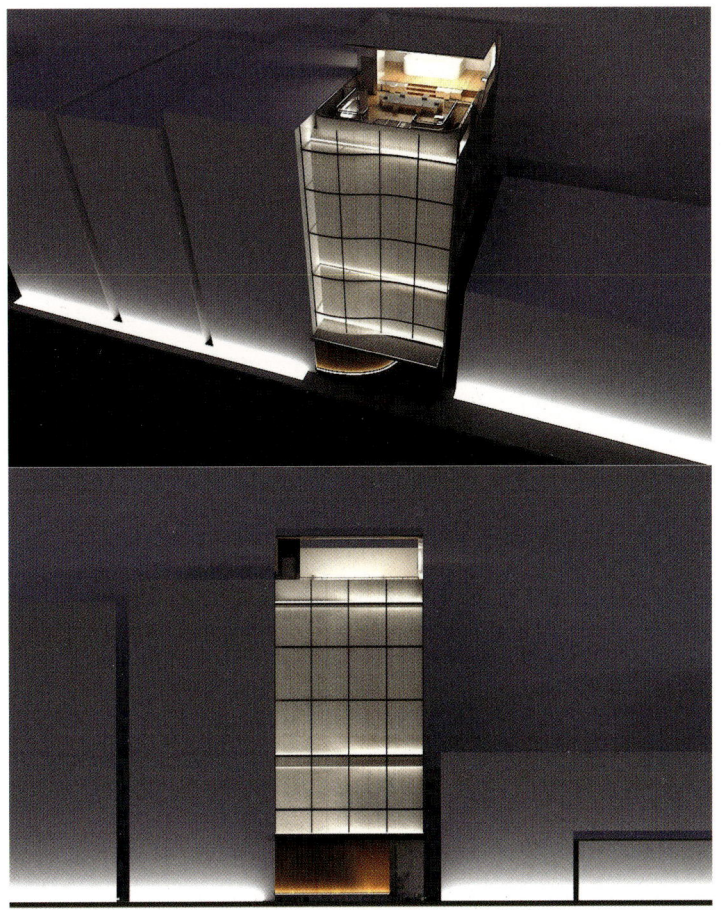

RENDERING

resemble the entryway of a cruise. Varnished wood that is commonly fitted on luxury yachts is used to lead patrons to the main entrance, the PAD is made with wooden-finished aluminium cladding for outdoor use, practicality and cost control.

Once the customers step foot in the vicinity, it is as if the journey of a luxury cruise ship sailing through the night sky is being told. Upon entering, customers are swept on a voyage that takes them from a bustling city to the calm body of water under the night sky, through the fog, light sea breeze, and the ocean waves.

The first floor holds the new experience all-day cafe and bar where customers can enjoy coffee or cocktails throughout the day while indulging in a night atmosphere. The use of dark-toned fittings and dark walnut along the arch profile creates the sensation of being on a luxury cruise ship. The copper

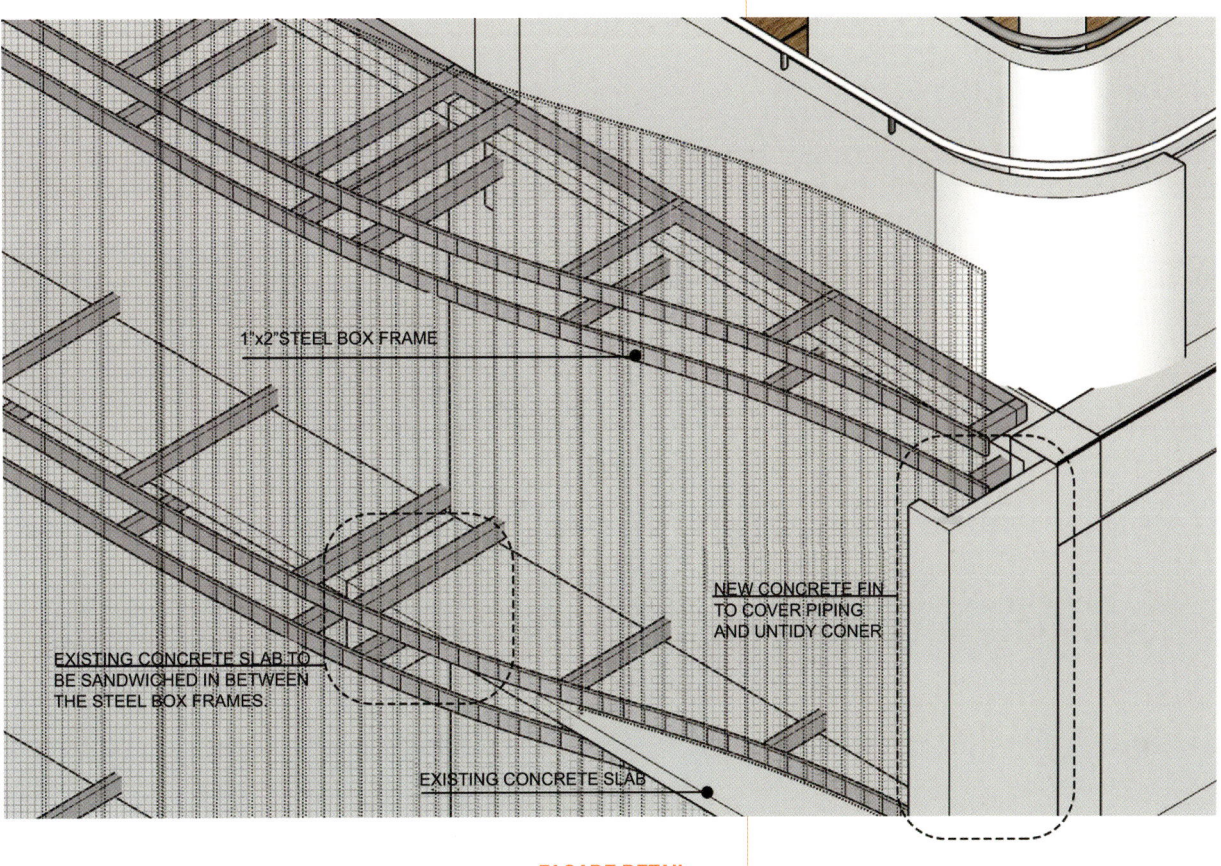

FACADE DETAIL

313

origami boat gleams through layers of hanging clouds as the main feature at the counter bar. A little twist on this floor is a hidden lab room that is designed to probe a secret under-the-ship lab where the distilling and tasting happens. The moon staircase, a curved staircase leading to the lab room, the mezzanine and the DJ booth, is yet another spot visitors will find hard to resist for a photo opportunity.

The second floor holds a luxury surf and turf. The concept of a luxury cruise ship continues throughout the design. This area encompasses a sense of a food cabin where customers dine by the cruise ship windows. The design focuses on retro-style cruising with the use of curves, arches, and round angles to create a sense of going back in time on a vintage cruise.

도시의 번잡함에서 벗어나 고급스러운 다이닝을 경험하다.
방콕 통로(Thonglor) 거리에 새로 들어선 레스토랑 Pennant는 통로 요리 체험을 위한 최초의 플래그십 매장 개발을 목적으로 세계적인 식음료 기업가들이 모여 설립한 것으로 고객에게 잊을 수 없는 경험을 선사한다.

콘셉트 디자인
수쿰빗(Sukhumvit) 메인 로드를 마주한 통로 거리 초입에 자리 잡은 기존 건물의 위치에서 우리는 바람에 날리는 돛을 닮은 곡선의 영감을 얻었다. 이는 수쿰빗 로드의 역동적이고 고객을 환영하는 반투명 곡선의 탄생으로 이어졌다. PAD가 있는1층의 외관은 2층과 대비를 이루도록 설계해 유람선 입구의 분위기를 연출했다. 주로 고급 요트에 장착되는 니스칠한 목재는 고객을 출입까지 안내하는 데 사용했으며, PAD는 야외용, 실용성, 비용 면을 고려해 목재 질감 알루미늄 외장으로 제작했다.
고객들은 이곳에 발을 내딛는 순간 마치 밤하늘을 항해하는 호화 유람선의 여정을

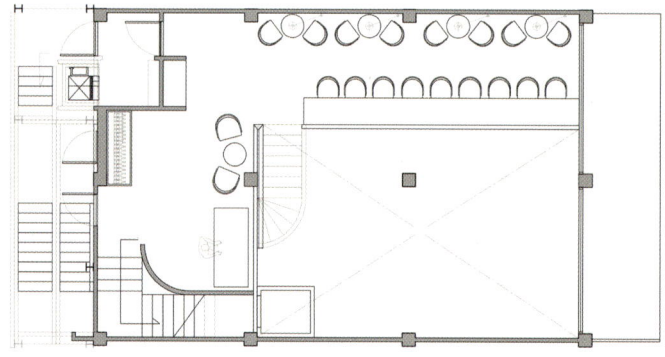

MEZZANINE PLAN

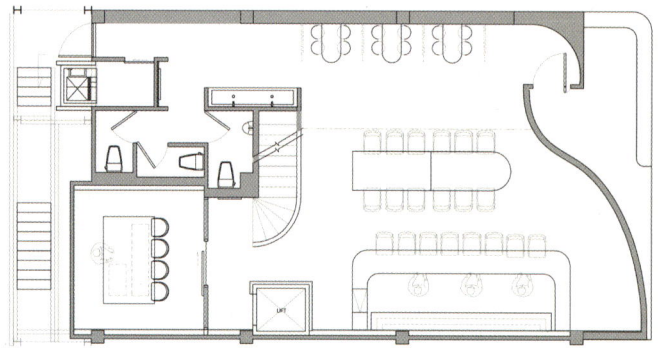

GROUND FLOOR PLAN

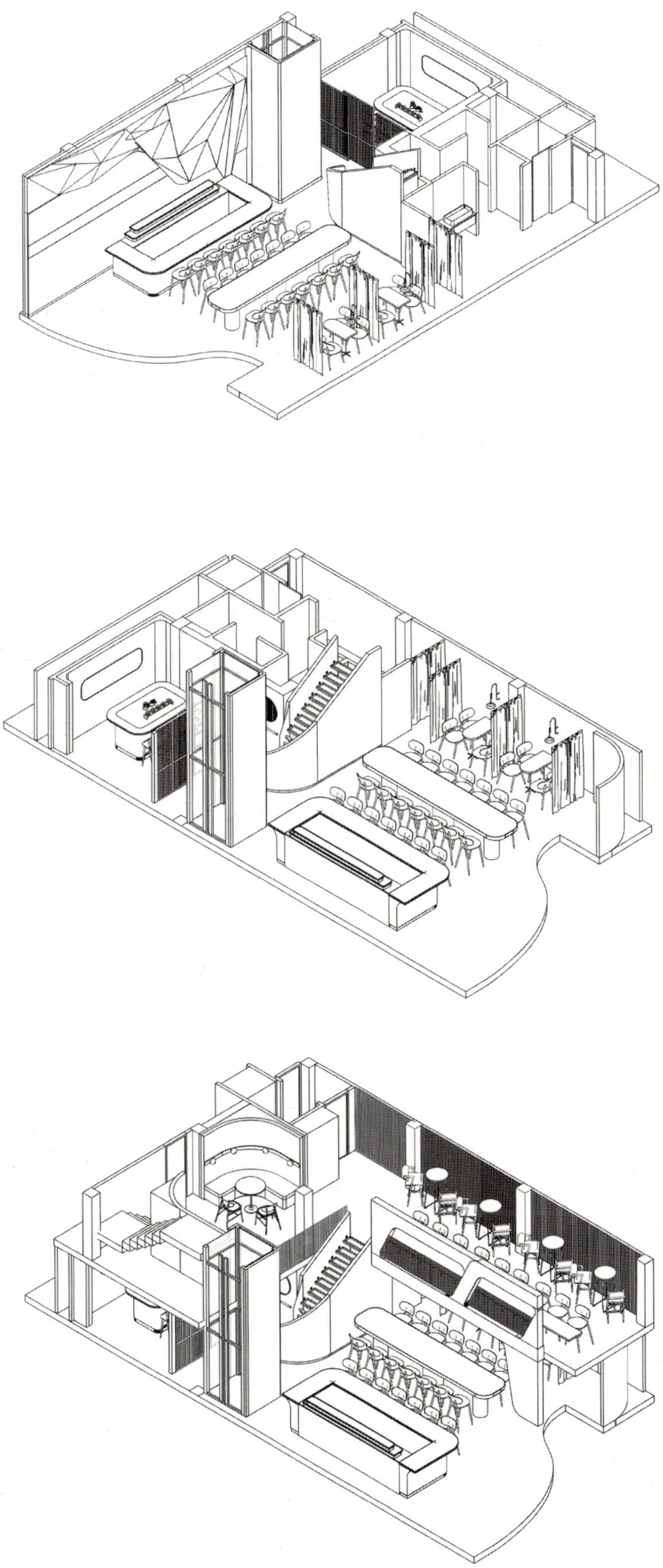

ISOMETRIC

315

INTERIOR MATERIAL

1. Laminate in Copper Finish : EDL
2. Wallpaper in Timber Pattern : Goodrich
3. Laminate in Timber Pattern : EDL
4. Black Marquina Marble : Stone Gallery
5. Wallpaper : Goodrich
6. Floor Hardener : By contractor
7. Poly carbonate : by contractor
8. Aluminium frame : by contractor
9. Bronze Mirror : In The Glass
10. Black Laminate : EDL
11. Rubber Floor : by contractor

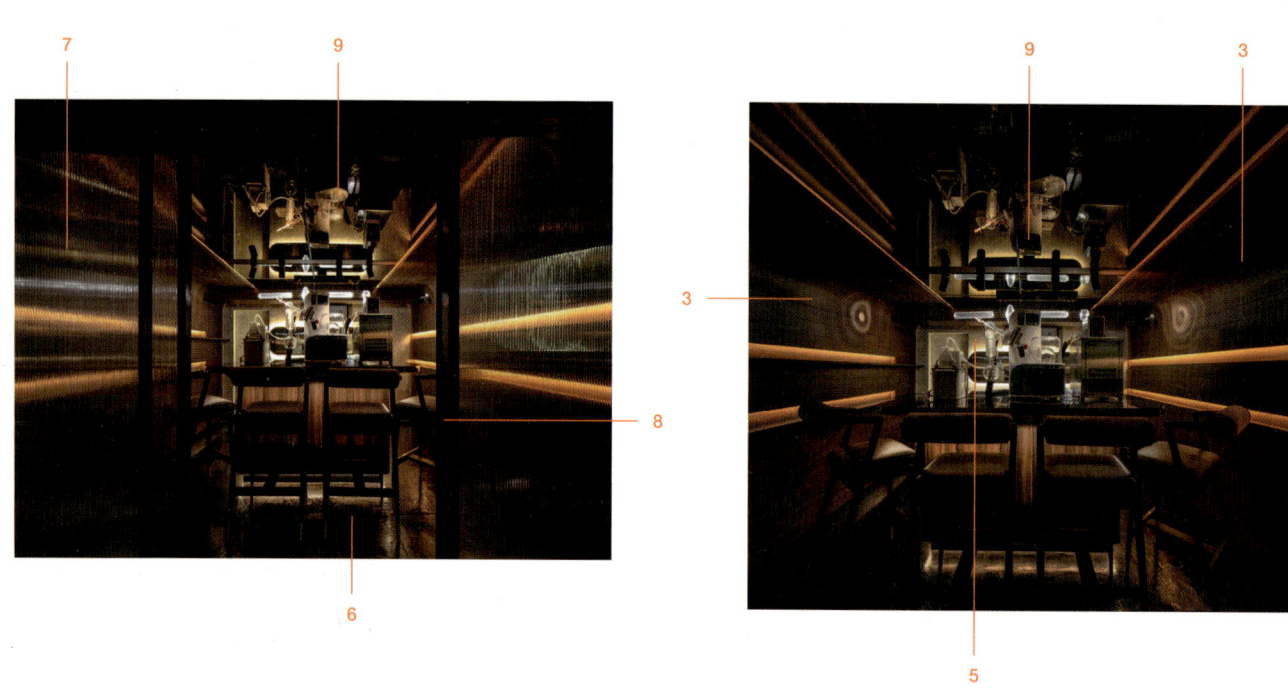

만끽하는 듯한 착각에 빠진다. 이곳으로 들어서는 순간 번잡한 도시를 벗어나 밤하늘의 고요한 바다에서 안개와 살랑살랑 불어오는 바닷바람과 파도를 헤치며 항해하는 듯한 느낌을 받는다.

1층에는 밤 분위기에 취해 하루 종일 커피와 칵테일을 즐길 수 있는 전에 없던 올데이 카페와 바가 마련돼 있다. 아치형 프로파일을 따라 어두운 톤의 비품과 다크 월넛 목재를 사용해 마치 호화로운 유람선을 타고 있는 듯한 느낌을 준다. 카운터 바의 주요 특징으로 구리로 된 오리가미 보트가 매달려 있는 구름 층층으로 빛을 발한다. 이 층의 작은 반전은 바로 숨겨진 실험실로, 증류와 시음이 이루어지는 선내 비밀 실험실을 탐험하도록 설계했다. 달의 계단은 실험실,

1 HIDDEN LED LIGHT
2 LED
3 ST-01
4 TOP EDGING IN STAINLESS
5 SS-01
6 SP-07

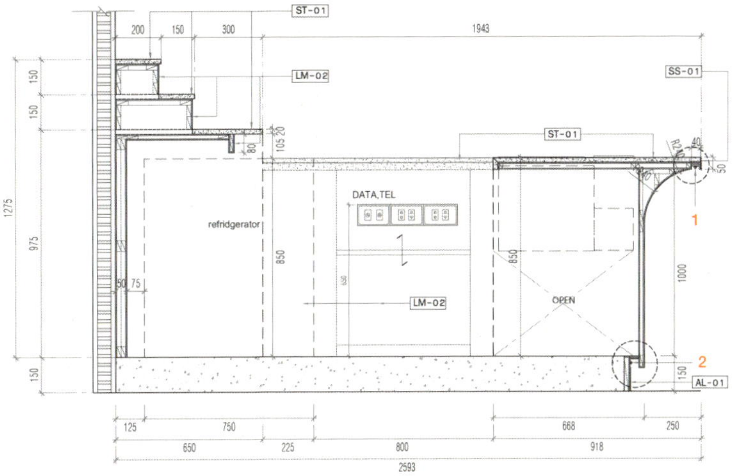

BAR SECTION

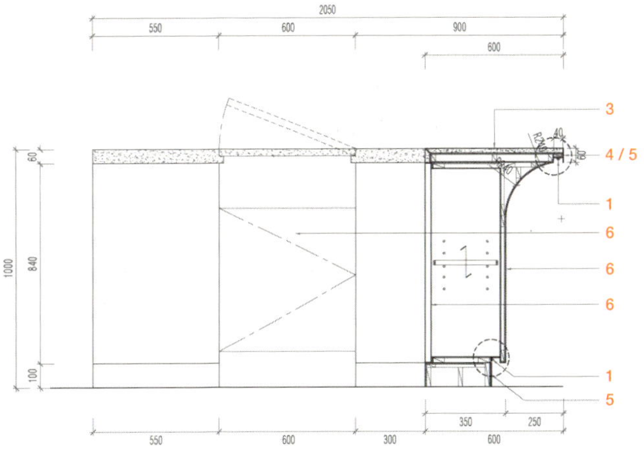

BAR DETAIL

복층, DJ 부스로 이어지는 곡선형 계단으로 방문객들이 지나칠 수 없는 포토 스팟이다.

2층에는 고급스러운 서핑장과 잔디밭이 마련되어 있다. 호화 유람선이라는 콘셉트는 디자인 전반에 걸쳐 이어진다. 이 공간은 유람선 창가에서 손님들이 식사를 하는 곳으로 푸드 캐빈이 연상된다. 곡선, 아치, 둥근 각을 사용한 복고풍 유람선에 초점을 두고 디자인하여 빈티지 유람선을 타고 과거로 시간 여행을 하는 듯한 느낌을 준다.

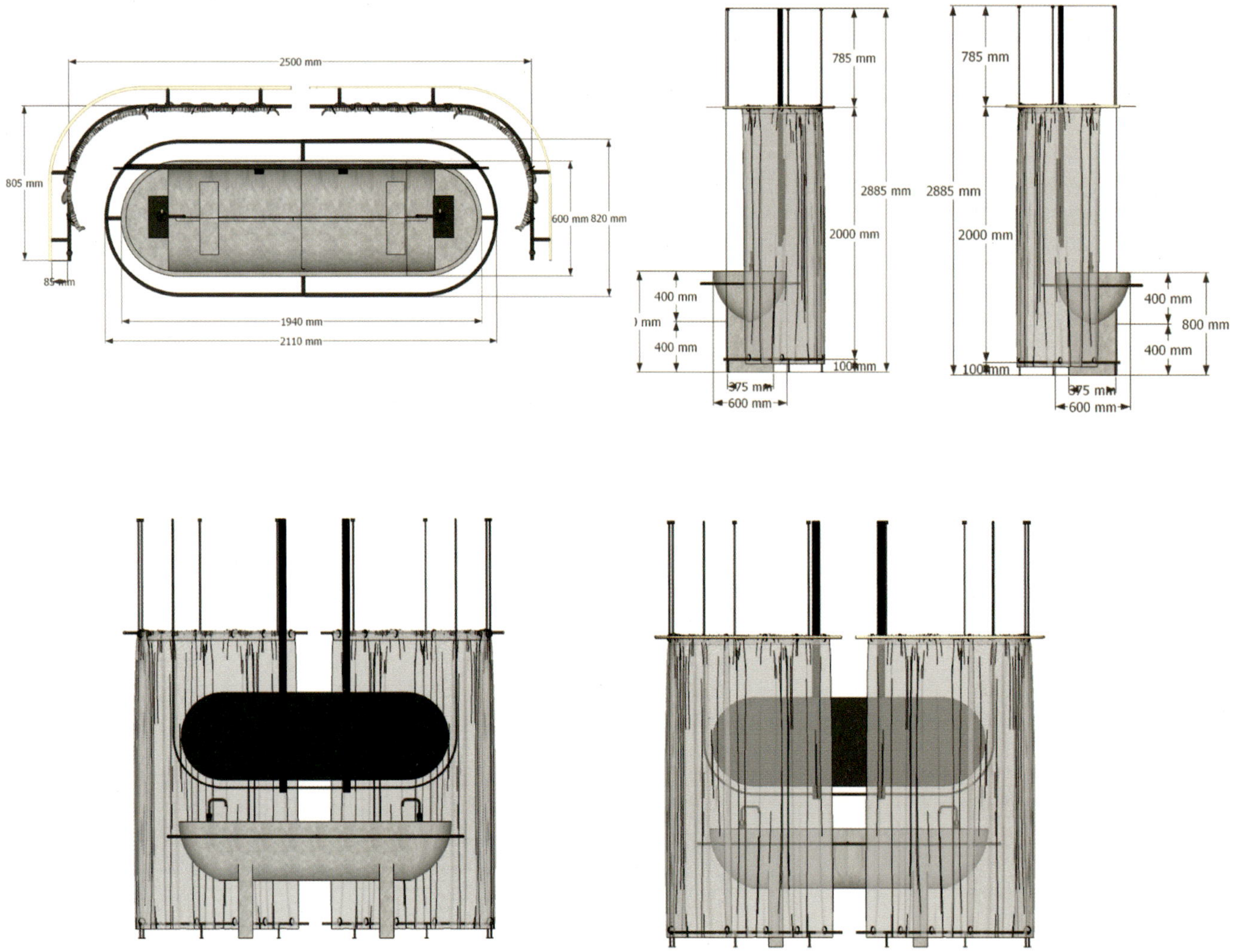

RESTROOM COUNTER DETAIL

Neighborhood Facility

PAD Space Artisan

PAD was established since 2015 for interior and landscape design practice based in Bangkok. We consider ourselves the space artisan. Our passion is bringing any kind of space to life with our expertise and craftsmanship. Our studio brings together a team of highly motivated, young and talented designers looking to make an impression in the local and regional markets.

>> www.padartisan.com

1. Black Laminate : EDL
2. Internal Paint : TOA
3. Tile (Terrazzo Pattern) : WDC
4. Tile (Black color) : WDC
5. Stainless Stee in Copper Finish : MAHA
6. Texture Glass : In The Glass
7. Curtain : Klazz
8. Black Marquina Marble : Stone Gallery
9. Alumimium Composit : SIAMBOND

VOL 7
근린생활시설 / NEIGHBORHOOD FACILITY

발행인 / 조배연
기획·편집 / 이선아, 국설희 (에뜰리에 www.etelier.kr)
디자인 / 아키랩(월간 건축문화)
마케팅 / 정순안

ISBN / 979-11-89659-47-9
출판등록번호 / 제2014-000167호

발행처 / 아키랩(월간 건축문화)
주소 / 서울시 서초구 양재천로13길 18(양재동)
전화 / 82-2-579-7747
이메일 / 1979anc@naver.com
정가 / 98,000원

Publisher / Cho Bae-yeon
Planning · Editing / Lee Sun-A, Kuk Seol-hee (E'telier www.etelier.kr)
Design / ARCHI-LAB (Monthly review of Architecture & Culture)
Foreign Business Dept. / Kevin Jung

ISBN / 979-11-89659-47-9
Registration No. / 2014-000167

Publishing Office / ARCHI-LAB (Monthly review of Architecture & Culture)
Address / 18, Yangjaecheon-ro 13-gil, Seocho-gu, Seoul, Republic of Korea
Tel / 82-2-579-7747
E-mail / 1979anc@naver.com
Price / USD 98

*저작권법에 의하여 보호를 받는 저작물이므로 어떤 형태로든 무단 전재와 무단 복제를 금합니다.

*All rights are reserved. Produced in Republic of Korea. No part of this book may be reproduced in any form without written permission of the publisher.